CITY OF FIRE

LOS ALAMOS AND THE BIRTH OF THE ATOMIC AGE 1943–1945

JAMES W. KUNETKA

Prentice-Hall, Inc., Englewood Cliffs, New Jersey

FOR MY FATHER

City of Fire: Los Alamos and the Birth of the Atomic Age, 1943–1945
by James W. Kunetka
Copyright © 1978 by James W. Kunetka
Photographs courtesy of the Los Alamos Scientific Laboratory

All rights reserved. No part of this book may be reproduced in any form or by any means, except for the inclusion of brief quotations in a review, without permission in writing from the publisher.
Printed in the United States of America
Prentice-Hall International, Inc., London/Prentice-Hall of Australia, Pty. Ltd., Sydney/
Prentice-Hall of Canada, Ltd., Toronto/Prentice-Hall of India Private Ltd., New Delhi/
Prentice-Hall of Japan, Inc., Tokyo/Prentice-Hall of Southeast Asia Pte. Ltd., Singapore/
Whitehall Books Limited, Wellington, New Zealand
10 9 8 7 6 5 4 3 2 1

Library of Congress Cataloging in Publication Data
Kunetka, James W   Date
  City of fire.
  Bibliography: p.
  Includes index.
  1. United States. Scientific Laboratory, Los Alamos, N.M. 2. Atomic bomb—History. 3. Los Alamos, N.M.—History. I. Title.
QC792.8.U6L674   1978   623.4'5119   77-16403
ISBN 0-13-134635-0

ACKNOWLEDGMENTS

Los Alamos is a place with a story that is long and rich. This history is concerned with only two years of its recent past. These years—1943 to 1945—witnessed the beginning of a scientific laboratory that took its name from the lonely New Mexican mesa it occupied. The impact of the work done at the Los Alamos Scientific Laboratory during this period extends far beyond the effort of the first atomic bombs built there. In two short years, Robert Oppenheimer, the first director, and his colleagues opened a new age, and none of us alive today can ever be very far from that mesa and all of its consequences.

There are many in Los Alamos who provided assistance and who spent time with me and shared their experiences and their interpretations of that time. I am grateful to Harold Agnew, the Laboratory's present and energetic director, and to John Manley and Norris Bradbury, two early leaders of the Los Alamos Laboratory. I also want to thank Arthur Schelberg, Berlyn Brixner, H. C. Paxton, Bill Stratton, Carson Mark, and Marge Dube, all members of the present staff.

The Laboratory's Information Services Division was most helpful and cooperated with my repeated requests for information and assistance. I am very grateful to Dell Sunberg, the division's director, and to Bill Regan, Faith Stevens, Art Freed, Dave Heinbach, Gilbert Ortiz, and Bill Richmond.

I am particularly grateful to three men at the Los Alamos Laboratory: Bob Masterson, Walt Bramlett, and Bob Krohn. Bob Masterson was very helpful as a liaison to many units of the Laboratory. Walt Bramlett sifted ancient files and stored documents for relevant material, and spent many long hours with me locked in a "secure" records vault. Bob Krohn was one of the first scientists at Los Alamos in 1943, and possesses an amazing grasp of the Laboratory's intricate history, as well as a wonderful sense of its importance in the modern era. He shared much with me, and reviewed this manuscript for security and for technical accuracy.

I also profited from the warm and fascinating afternoons with Dorothy McKibben, and from the visits with Barbara Storms, Ray and Elizabeth Gray, Malcolm and Emma Knowles, Peggy Pond Church, Priscilla Green Duffield, and Margaret Wohlberg.

At the U.S. Army's White Sands Missile Range, I want to thank Charles Poisall, who spent two days with me and who shared a long trip into the desert to visit Trinity Site and to witness the expanse of the Jornado del Muerto.

At the Energy Research and Development Administration (formerly the Atomic Energy Commission), I owe thanks to Charles Marshall, the retired director of the Classification Division, Melvin Neff, and Richard Hewlett.

At the National Archives, I am very grateful to Edward Reese, a talented researcher, and the most knowledgeable person on the vast holdings of the Manhattan Engineering District.

Others to whom I owe thanks for encouragement and assistance are Sherry Kafka Wagner, Warren Bayless, Oscar Collier, Marge Heath, Elizabeth Davis, Dennis L. Boyles, George A. Sheets, Donna Pfefferle Niece, and Martha L. Smith.

I am certainly grateful to three Los Alamos residents who housed me during my many visits and who showed a continuing interest in this work: Bob, Jeannene, and Rob Masterson.

November 1977

CONTENTS

PART ONE: ALPHA

    1. City of Fire   *3*
    2. Origins   *10*
    3. World in Change   *20*
    4. The New Alliance   *32*
    5. Laboratory on a Hill   *43*
    6. The Actors   *57*
    7. Organization, 1943   *64*
    8. The First Year: 1943-1944   *74*
    9. Life on the Hill   *90*
    10. The Second Year, 1944-1945, Part I   *109*
    11. The Second Year, 1944-1945, Part II   *116*
    12. The Harvest   *131*

PART TWO: OMEGA

    13. Homestretch   *141*
    14. Dawn   *158*
    15. Japan   *174*
    16. Change of Guard   *188*

PART THREE: RETROSPECTION

    17. Retrospection: 1978   *207*

NOTES   *214*
SOURCES & BIBLIOGRAPHY   *220*
GLOSSARY   *224*
INDEX   *229*

PART ONE
ALPHA

SOME SAY THE WORLD WILL END IN FIRE
Robert Frost

1. CITY OF FIRE

July 15, 1945. Sunday afternoon.

Fat Man lay quiescent on its special cradle a hundred feet off the ground. It was sheltered on its tower by a small cabin with three walls of metal and one of canvas. Small teams of men entered for last-minute checks and adjustments and left through a small trapdoor in the wooden floor. The bulky object that dominated the room was a dark metal sphere nearly five feet in diameter with wire cables like thick tentacles sprouting from its body, leading to aluminum boxes mounted on the cabin floor. For the moment the device rested quietly, disconnected from its electrical source ten miles away. Only the wind stirring the canvas flap and the grinding sound of an occasional truck broke the stillness of the New Mexico desert.

Fat Man was the name given by scientists to the first atomic bomb. In a dozen hours, the bomb would be asked to explode and to confirm nearly three years of work and over two billion dollars in expenditures.

4/ Alpha

Fat Man was the child of Los Alamos, a city and scientific laboratory hidden in northern New Mexico. Remote, almost inaccessible, Los Alamos was a name known only to the residents of the town and nearby neighbors. Except in the minds and files of a few dozen military men, political leaders in Washington, and scientists in scattered laboratories across the country, the Los Alamos Laboratory did not officially exist. The design and production of atomic weapons was to be one of the great secrets of World War II.

Los Alamos was a community of hastily constructed homes built around a complex of scientific laboratories and offices; the small city lay secluded in the splendid beauty of the Jemez Mountain chain on a mesa that was one of the geologic fingers of Pajarito Plateau. The closest city of any size was Santa Fe, the ancient capital of New Mexico, some 40 miles to the southeast.

For over two years—since early 1943—the Laboratory had steadily grown from a few dozen men to a complex of over 5,000 scientists and technicians. This once slumberous mountain land formerly occupied by a private school for boys was the center of a furious effort to beat German scientists in the construction of the first atomic weapon. The work grew out of nearly two decades of scientific investigation that had ultimately revealed the great energy that lay trapped in the atom's tiny nucleus. The challenge of tapping that power had been taken up by a remarkable coterie of American and European scientists who had pursued the discovery relentless. July 15, 1945, was the day set to test several years of exhaustive work.

Most of the residents of Los Alamos had been brought to the Mesa by a theoretical physicist named Robert Oppenheimer. For the last two years, Oppenheimer had directed the most unusual scientific laboratory in the world and as director, he now guided the critical arm of the government's vast Manhattan Project. Together with a testy Army general, he had brought the world to the edge of a new era.

Los Alamos had finally fashioned a billion dollars' worth of plutonium and uranium into the hearts of two atomic bombs. One of these—the plutonium Fat Man—employed an astonishing array of scientific discoveries and theories in a metal sphere composed of layers of carefully shaped explosive charges around a core of a previously unknown, artificially created new element. Fat Man now rested on its tower two hundred miles to the south of Los Alamos. Its inventors hoped it would explode in the desert with a brilliance greater than that of the sun.

Oppenheimer, more than anyone else, had led Los Alamos to the coming test. Always small and frail, he was now gaunt and his weight had fallen to less than 110 pounds. Two years of leading his colleagues and of ameliorating differences between the Army and scientists had taken its toll; the long hours of work and tension showed in his appearance. The mountains and air of New Mexico had in the beginning been his source of pleasure and continuing energy. Now he seemed near exhaustion. And after two years of work he was ready to have Monday's test prove the value of his direction. The bomb would either work or fail, and with it, his reputation as scientific leader would blaze or wane.

Oppenheimer had been born to wealthy New York parents. Intellectually able, he moved easily through Harvard and two European universities to earn his doctorate before he was twenty-three. At Stanford and the University of California at Berkeley, he was regarded as a brilliant and charismatic teacher. His grasp of physics and leadership ability earned him an invitation to be part of the fledgling movement that led to the unraveling of the secrets of atomic fission. In an odd match of personalities, he had been given leadership of Los Alamos by a hard-working but difficult Army general named Leslie Groves.

Major General Groves was the epitome of the career Army officer. A graduate of West Point and builder of the Pentagon, he had served in major engineering and construction projects around the world as part of the Army's Corps of Engineers. He was efficient and demanding and found it hard to understand civilian life, much less the largesse of academic communities. Where Oppenheimer was lean, with the refined features of an aesthete, Groves was corpulent in double-starched uniforms that constrained his bulk. Groves always sought to appear well tailored and groomed, and sported a neatly trimmed mustache. Oppenheimer was attired in expensive clothes that always appeared too large for his frame. He was an intellectual with a casual manner, who punctuated his conversations with continual gesturings of a pipe or cigarette.

Groves had been handpicked by U.S. chief of staff General George Marshall in 1942 to head a mysterious new consortium of the Army and university laboratories called the Manhattan District Project. Its purpose was to produce large quantities of nearly-pure elements called Uranium 235 and Plutonium 239. These elements were fissionable, or capable of a chain reaction, and under the right circumstances it was hoped they could be made to produce an explosion that would rival the force of an earthquake or volcano.

In 1943, however, only microscopic amounts of Uranium 235 existed, and plutonium could be seen only through a microscope. Groves was charged with building and managing huge industrial plants to produce the needed quantities of both elements and to oversee the design of a weapon capable of causing a nuclear explosion. In 1942, no one could be sure whether the Germans would beat America to it.

The complexities of producing such a weapon—even if plutonium and uranium *were* available in quantity—were monstrous. Scientists had only achieved the fission process in 1939. While both German and American scientists immediately recognized the implications of fission in the creation of a weapon, it took men of international prominence like Albert Einstein and Leo Szilard to interpret the concept and explain its potential military application to President Roosevelt and American military officials. Nearly four years had been spent in fission experiments at widely scattered universities before Robert Oppenheimer and others pressed for a consolidation of both theoretical and applied work within a single laboratory. Groves had been impressed with Oppenheimer's arguments, and eventually offered him the directorship. Together, the men had chosen New Mexico as a site for the new center. And although it was not easy at first, Oppenheimer eventually succeeded in attracting men like Enrico Fermi, Edward Teller, Hans Bethe, and Robert Bacher to the new laboratory. Los Alamos quickly became a distinguished research center hidden in the mountains.

Now, in July 1945, Oppenheimer and his colleagues had two bombs ready, one made from plutonium and the other made from uranium. The question remained: Would they work? The uranium bomb was simple and straightforward. Constructed like a gun, it fired one piece of uranium into another at tremendous speeds. Among the Laboratory staff, there was a sense of confidence about the uranium bomb. In fact, a team of men was already preparing the weapon for shipment to a secret island base in the Pacific for quick use against Japan. Only the plutonium bomb seemed uncertain. It employed a complex assembly of explosive charges arranged in concentric layers around a plutonium sphere. These fragile components had been designed according to theories developed by Los Alamos scientists. Would these parts actually combine at the last second to release energy in the form of a powerful explosion? Only a test of the prototype weapon would tell, and for nearly half a year scientists and technicians had sweltered in the desert to prepare for just such a test.

The land around the test site was named the Jornado del Muerte by early Spanish travelers: the Dead Man's Route. And to the new visitors from Los Alamos, it seemed that death had found its permanent home. For months, Laboratory trucks and cars made the laborious descent from Los Alamos to Santa Fe, then Albuquerque, and finally to an unmarked cutoff that led into the heart of the Jornado. Hardly anything grew here except low shrubs and desert flora blasted by the merciless sun. Even small creatures like scorpions and lizards hid when the sun was at its fiercest.

But Los Alamos had chosen the Jornado precisely for its isolation and scarcity of life. Oppenheimer had nicknamed the site Trinity, after a line in a poem by John Donne. The land was borrowed from the Army's Alamogordo Bombing Range, and several hundred square miles were placed off limits to the B-29 crews who regularly practiced bombing runs on its wide expanse before assignment overseas.

Trinity was conveniently isolated; the nearest town was thirty miles away. In a particularly flat run of land, flanked to the east by the Oscurro Mountains, the Laboratory built a base camp, and ten miles away built its tower for Fat Man. The center of the tower had been named Ground Zero. Emanating from the towerlike spokes of a giant wheel were roads leading to three major observation towers and control shelters, each 10,000 yards away. These shelters would control instrumentation created and arranged across the desert for a thorough visual and electronic record of the explosion. Groves, Oppenheimer, and other senior scientists would gather at the South Shelter to assume overall command of the test.

By July 15, both the scientists and the desert were ready. Over a thousand miles of wire laced the desert floor connecting instruments and gauges to command centers. Communication lines dropped lazily in the sun from man-high poles connecting each shelter with observation posts and with Base Camp. Fewer than fifty of the hundreds of miles of dirt roads criss-crossing the land were paved, because General Groves had suddenly felt cost-conscious and objected to the expense of blacktopping every mile. Dotting the desert floor were observation posts built like half-tents where a single man or small teams with specialized instruments or experiments could sit at "Zero Hour."

Throughout the day, dignitaries arrived at Base Camp from Albuquerque. From the West Coast came Ernest Lawrence, builder of the giant cyclotrons and a Nobel Prize winner for his early work in physics. From the East came James Conant, president of Harvard University, and Vannevar Bush, director of the powerful Office of Scientific Research and Development. Joining General Groves were Brigadier General Thomas Farrell and military leaders from other laboratories within the Manhattan Project.

By Sunday afternoon, the peace at Trinity slackened considerably. Most equipment was ready, and only a few scientists were still busy with last-minute adjustments or cantankerous equipment. Fat Man had been partially assembled on Saturday in a small stone ranch house a few miles from Ground Zero. The team of men putting Fat Man together for the last time at the base of the tower was led by Marshall Holloway. Very delicately he had dropped the last portion of the plutonium sphere into the core of the bomb. The piece stuck momentarily and then slid into place. He bolted the remaining pieces of the metal shell together. Slowly the bomb had been raised to the top of the tower and through its trapdoor in the platform. Once it was in place, another team of men connected the firing cables, leaving the final arming of the bomb to Kenneth Bainbridge, the overall director of the Trinity test. Only after consultation with Groves and Oppenheimer would Bainbridge lock the bomb into its final position. Once this was done, control would shift to the South Shelter for ignition by a battery of electronic switches.

The weather seemed uncooperative. The Trinity meteorologist had predicted rain and the test was rescheduled from Sunday night to Monday morning. Early in the evening there was a distant rumbling in the mountains from a summer storm, and within an hour a light rain fell and lightning etched the summer sky. When the weather failed to clear, the test was again rescheduled, this time for 5:30 A.M. One man was particularly uncomfortable. Donald Hornig, a 24-year-old explosives expert, had volunteered to baby-sit with the bomb until early morning. All through the rain and lightning Hornig remained huddled in a corner of the tower's metal cabin next to nearly thirteen and a half pounds of plutonium and five thousand pounds of high explosives. His only diversion came from an adventure novel set in the South Seas.

At Base Camp, the night chill accentuated doubts and fears. Oppenheimer was unable to sleep and paced from building to building until General Groves ordered him to his tent. Groves was not without his own uncertainties. He had ordered several news releases prepared in advance of the test to cover explosions of several magnitudes; one would not be needed in the case of a complete failure. How Groves would explain a dud costing two billion dollars to the Army and to President Truman he did not know.

Several thousand miles away, in a bombed suburb of Berlin, President Harry Truman was preparing for his first meeting with Churchill and Stalin. He had been told about the new weapon only hours after taking office; it was a complete surprise to him. Roosevelt had never confided in him about Los Alamos and the Manhattan Project. Now, three months later, he was hoping that success in New Mexico would mean a swift end to the war and some political leverage with Stalin and the new leaders of the postwar world. A careful code had been worked out by Groves to alert his secretary in Washington on the results of the Trinity test. Good word was to be sent immediately to Truman's military aides in Germany.

Everyone waited for the weather to clear and for their work to bring an early dawn to the desert.

In Los Alamos there were expectations and uneasy stirrings. For weeks men had been leaving their families and disappearing to some secret spot south of the Mesa. Bits of sand found in the cuffs and pockets of returning men provided clues to the location.

There was still important work in Los Alamos to be done: both a uranium bomb and another Fat Man were being prepared for shipment overseas. The technicians in the concrete plutonium purification plant still labored to refine the precious substance to its pure metallic state. Others, in buildings located in the canyons below Los Alamos, continued to fashion explosive charges into small pyramids that when ignited would compress plutonium into itself and trigger a nuclear explosion.

Like Trinity, Los Alamos settled restlessly into Sunday evening. Some wives and children were camping out in the mountains that overlooked southern New Mexico, in hopes of catching a glimpse of the explosion. Hot steam from smokestacks in the laboratories and generating plants billowed up to the sky in clouds. For over two years the mountain city had trembled as men probed and stoked nature's elemental forces. Soon, two hundred miles away, these individuals would become men of fire.

2. ORIGINS

It was Robert Oppenheimer who brought the U.S. Army, and eventually the world, to Los Alamos. In October 1942, he rode in an unmarked car with three military companions along a little-used New Mexico highway. The four men had just left Jemez Springs, a small valley at the edge of the Jemez Mountain chain. The highway twisted its way up into the mountains where fall had already begun to make its changes of color in the tall aspen trees.

The senior man among them was Major General Leslie Groves of the Corps of Engineers. Groves had only weeks before stepped into command of a new and highly secret government organization called the Manhattan Project. With Groves were another Army officer, Lieutenant Colonel William Dudley, and an enlisted Army driver from a military base in Albuquerque. The conversation centered—to the enlisted man at least—on the suitability of the geography for some sort of new Army installation.

The automobile headed toward a level plateau in the mountains inhabited by the owner and students of a private boys' school.

Highway Four was the back entrance to Pajarito Plateau; it meandered gracefully through the mountains into the region where the aspen were joined by juniper and piñon trees. At one point, to the left of them, the mountains split into another valley where ancient Indian cliff dwellings dotted the soft pumice cliffs. Then unexpectedly, the road turned into the sprawling complex of log homes and school buildings.

It was a fateful day. A few young boys turned to watch the car and its occupants with curiosity; they seldom had visitors in their lonely mesa. It was an interesting sight: only one of the men wore civilian clothes and he was small and thin, as compared to the others. Oppenheimer, however, was no foreigner to the mesa. For years he had ridden horseback over Pajarito from his small ranch near Santa Fe in the Pecos Mountains. Several times he had crossed over mountain trails to Los Alamos Mesa.

The Army men hardly noticed the students; instead, they studied the terrain and buildings and referred to maps and papers. Oppenheimer shared more of a naturalist's love for the land, and it was he who had suggested to Groves that Los Alamos might be an excellent and clandestine setting for a small scientific laboratory. Scientists and their families could live and work in an atmosphere that was confined but exhilarating.

General Groves was not impressed with the scenery. He was a mechanical engineer with years of Army experience in construction in all sorts of places. He was concerned with pragmatic questions: How much would it take in labor and money to convert this mountainous landscape to a military post with a scientific laboratory as its core? Would it be possible to quarter scientists here, and to keep the development of an atomic weapon a secret?

The Los Alamos that Groves and Oppenheimer saw that day had its own long and colorful history before their visit. The contemporary geography was the result of birth by fire and violence. Volcanos erupted again and again over millions of years to form the Jemez Mountain chain. In the midst of these mountains, between two other chains—the Sierra Nacimientos and the Sangre de Cristo—lies a huge geologic circle. Formed like an eye, this is made up of the aged remnants of one singular volcano that exploded and collapsed upon itself to form a basin 20 miles across. Over the aeons, the basin, or caldera, formed a lake that wore down parts of its rim to drain away to what is now the Jemez River gorge.

For the Indians, the name Jemez meant Place of the Boiling

Springs. Later, with the Spanish, the lush, grassy valleys took romantic and colorful names like Valle Grande, Santa Rosa, and Jaramillo. On one side of the caldera's rim is an irregular plateau that extends along and halfway up the eastern slope of the Jemez Range to overlook the Rio Grande River.

The Spanish called the plateau Pajarito, or Little Bird. It spreads eastward from the Jemez foothills and rises and falls several thousand feet to resemble a large spreading hand, with each finger forming a mesa. Between the fingers are deep, vertical canyons whose precipitous cliffs are like multicolored paintings set in the eroded volcanic earth. Behind the plateau are the dramatic Jemez Mountains, which rise over ten thousand feet. The peaks are overrun with a caravan of fir, aspen, and other tall trees. Early in the morning, or after a rain, the air is rich with the scent of piñon and juniper.

Men came to New Mexico thirty thousand years ago. At first, perhaps, they were nomads, roving hunters with few tools and armed with simple spears thrown from hand-held devices called atlatls. Only later, after thousands of years, were small settlements begun. By 1000 A.D., the Indians were regularly farming by day and taking to small caves in the cliffs for protection at night.

All around Los Alamos, in lonely canyons or on top of mesas, small clusters of Indians grew and prospered. Bandalier, Puye, Tsankawi, Otowi, and Tshirege: these were the Indian names of pueblos that flourished in New Mexico while Europe was in its Dark Ages.

With the advancement of agricultural skills, living centers spread throughout the region. Just below Frijoles Mesa, in its canyon, the pueblo and cliff dwellings of Tyuonyi were built. At its zenith, Tyuonyi contained a vast circular community house with dozens of rooms and three large kivas, or ceremonial rooms, built into the ground. The region's population increased because Indians from what is now called the Four Corners area—the portions of New Mexico, Colorado, Utah, and Arizona that meet—came and settled the Canyon in the thirteenth and fourteenth centuries. Groups of extended families enlarged natural recesses in the soft walls of the canyon and decorated them with designs and petroglyphs that resemble modern-day scientific calculations.

Only a few miles away from Tyuonyi was the pueblo of Tsankawi. Built in the twelfth century, it was continually occupied for more than three hundred years. At its largest, the community buildings rose two and three stories from its perch on the edge of a canyon. To the north, the Puye Indians built another cliff city that spread from its mesa perch into the walls of the canyons below.

These early residents became known as the Keres people—loose-knit tribes of Indians that had come from the Zuni-Acoma area of western New Mexico. Much of their architecture was characterized by large, massive community buildings with many small rooms or chambers. About the year 1300, persistent droughts to the North drove Indians from Mesa Verde and Chaco Canyon into the area. These peoples were called the Tewa—a name that characterizes their language today—and they, too, built pueblo cities that lined the rims of the mesas. They gave their cities the names of natural events or objects such as Te'ewl (little cottonwood gap), Yunque-yunque (down the mockingbird place), and Pininikangwi (dwarf-cornmeal gap). Their major living centers were built at Puye and Tshirege and rose like fortresses to mirror their ancient homes like the sky-city of Acoma. The single pueblo of Tshirege commanded a magnificent view of the Jemez canyons and contained over 600 rooms.

All of these towns were eventually abandoned. The prolonged droughts that had brought the Tewa to the area intensified and drove them away. The Indians at Puye migrated to what is now the Santa Clara Pueblo; the people of Otowi and Tsankawo moved to San Idelfonso. By the time of the arrival of the Spanish, the Pajarito pueblos were deserted.

Europeans arrived in midcentury to find the plateau empty. In 1540, Francisco Vasquez de Coronado made his way from Mexico into the Rio Grande Valley. Coronado and his men sought the seven golden cities of Cibola and found only the peaceful and impoverished descendants of the Keres and the Tewa. Forty years later, in 1581, another Spanish exploration made its way through the valley. This journey was led by a Franciscan lay brother named Augustin Rodriguez. His visit was peaceful and for God; he left the conquest of the land to Don Juan Oñate, who followed him several years later.

Oñate was a wealthy mine owner from Zacatecas in Mexico. He traveled at his own expense to colonize new lands north of Mexico. After numerous delays, he arrived with an army of soldiers and settlers that numbered 400. It was February 7, 1598, and on April 30 Oñate declared New Mexico a part of Spain. From what is now El Paso, Texas, he had made his way north and established the first Spanish seat of government at Yunque-yunque.

Oñate was only the first movement of the Spanish conquest: following him were priests and merchants, drifters and violent men. Oñate was soon replaced by a harsh man known as Peralta, whose history included a charge by Spain of crimes against the government. The charges included a refusal to obey Royal decrees, mistreatment of friars and Indians, and punishing the Acoma and Jumano Indians with "especial" cruelty. As a people, however, the Spanish quickly enough settled and married among the Indians. Gradually, whole tribes became Christianized.

The Spanish would not keep the Valley and New Mexico for long. Mexican independence replaced rule by Spain, and Mexican rule was replaced by General Stephen W. Kearny, Commander of the U.S. Army of the West. In 1846 he entered New Mexico and commandeered the territory without force. Two years later, the Treaty of Guadalupe-Hidalgo legitimized the secession of the New Mexico territories from Mexico to the United States.

One Civil War battle fought at Glorieta Pass permanently won the western portion of the continent for the North. With the end of the war, railroads soon knifed through the land and opened it to settlers, miners, ranchers, and exploiters from the East and West. On January 6, 1912, the territory of New Mexico was formally admitted as the 47th state of the Union.

By the turn of this century, most of the Plateau had been settled by homesteaders. A wealthy Swiss named Adolph Bandelier began several years of explorations in the ruins of Frijoles Canyon. His novel, *The Delight Makers,* romanticized the lives of prehistoric cliff dwellers in Frijoles and first brought the attention of the world to the lonely Plateau. Soon, guest ranches for wealthy tourists were opened in the canyons of Pajarito. A newly married couple, George and Evelyn Frey, established one ranch that could be reached only by horse; mail and supplies were brought down by a small cable car. For a few years their ranch even sported a French chef.

A wealthy Detroit businessman named Ashley Pond also opened a "dude" ranch for automobile magnates from the East and their clients. Pond located his ranch in Frijoles and offered his visitors an exhilarating climate and abundant hunting, fishing, and horseback riding. Just a few years before, Pond had attempted to open a private school for boys in a nearby canyon, but before the opening the new school was destroyed by a flash flood.

The eventual success of the dude ranch, however, never diminished Pond's hope of starting a private school. His own love for the area came from his childhood: he had been a sickly child and had been sent by his parents to New Mexico to recuperate. A rugged life in the outdoors in the company of lumberjacks had revitalized his health and his future. He began to envision a school for boys like himself—a school that would combine a sound scholastic regimen with the rustic purity of the outdoors.

With inherited money, Pond finally began his second school

in 1916. At first it was little more than a few log buildings. Slowly he added a main house, stables, a barn, and more buildings for students and faculty. He was fortunate to hire Arthur J. Connell, a former New Yorker turned forest ranger and Boy Scout leader. Connell, like Pond, had come to New Mexico on a visit. He fell in love at once with the air and mountains and stayed. He was not an educator, but shared Pond's love of outdoor living and the belief that rigorous training was linked to sound education. Connell in turn hired the school's first faculty member from Yale and gave him the role of Headmaster and responsibility for developing the school's curriculum. Within a few years, other Ivy League graduates came as teachers, and the school offered course work in English, Latin, French, Spanish, mathematics, history, several sciences, and art. The school also had mandatory athletics that included horsemanship and camping.

Pond and Connell left the Headmaster to schoolwork, while they devoted themselves to developing the boys' physical and spiritual well-being. Pond eventually turned ownership of the school over to a board of directors. He served as a member of the board until his death in 1933.

Each boy wore a uniform patterned after that of the Boy Scouts—the school was officially chartered as New Mexico Troop 22—and woke every morning at 6:30 to exercises and a hearty breakfast. Each boy was assigned to one of four groups, organized by physical maturity rather than age. The groups were named after trees: piñon (the smallest boys), juniper, fir, and spruce. In 1916, Connell had to recruit the first student, Lancelot Inglesby Pelly, the son of a British Consul stationed in the United States. Despite the small numbers of the early student body and the high tuition ($2,400 a year), the school grew and reached its peak in the 1930's with 45 students plus staff and families. Graduates of the school fared well: most went on to become successful lawyers, doctors, and corporation presidents.*

By 1942, the school had met and exceeded all of Pond's original expectations. Only the war seemed to disrupt the school's success: young teachers were being drafted and it was becoming increasingly difficult for Connell to recruit the needed faculty. For the students, however, the visit by Groves and Oppenheimer in November seemed no more unusual than the airplanes that soon began to fly low again and again over the mesa.

---

*Students of the school included author Gore Vidal; Arthur Wood, president of Sears, Roebuck and Company; William Veeck, columnist and sportsman; John S. Reed, president of the Santa Fe Railroad and associated industries; David Osborn, director and playwright; Roy Chapin, Jr., chairman of the board of American Motors Corporation; and John Crosby, founder and director of the Santa Fe Opera.

16/ Alpha

Leslie Groves felt himself to be under great pressure. America had been in the war now for almost a year. Deep in Russia, the German troops were holding out against a renewed Russian attack on the Stalingrad front. American and British Expeditionary Forces had just landed in French North Africa. In the Pacific, American Marines had recaptured Guadalcanal, and naval forces were preparing for a crucial battle in the Solomon Islands. Just a few weeks before, on September 17, Groves had been handed the biggest task of his life. Recent discoveries in physics suggested that under the right conditions, a new and powerful bomb could be made to explode with less than a handful of purified uranium. This possibility emerged from a startling series of scientific discoveries that the Americans and British had every reason to believe was under intense study by the Nazis. Groves had been ordered to direct a new and powerful alliance between the Army and a scattered array of scientific projects studying uranium. This new venture had been given the code name Manhattan Engineering District, or more simply, the Manhattan Project.

Groves' lean companion on the mesa was a scientist of some reputation who recently had led his colleagues in urging a consolidation of all uranium projects under the direction of one laboratory. An atomic bomb was theoretically possible, and Oppenheimer suggested that science now needed the time and means to perfect the technology to cause the explosion. What was needed, he urged, was the creation of a special weapons laboratory. Groves had agreed, and the search for the right location ended on the mesa.

Groves made his decision: Los Alamos was it. The mesa met the selection criteria he and Oppenheimer had set some weeks before. Isolation was the most important requirement for Groves. Security leaks in a Chicago uranium project had convinced him of the need for tight security.

Both men were looking for other things as well. In addition to isolation, the prospective site needed adequate space for laboratory buildings and for testing. No one thought that an actual bomb test would be conducted in proximity to laboratories, so adequate rather than unlimited space would serve the new project. But a favorable climate would also be necessary to permit year-round work, as well as regular access by roads and railway. Groves' Engineering Corps experience was useful. The project needed to have construction materials readily available in necessary quantities, as well as sufficient water and power. The general also wanted civilian populations to be at a distance. He didn't want scientists—whom he regarded as eccentric children—to be tempted by proximity to major living centers. Preferably, the site would be remote enough from the East and West coasts to act as a deterrent to possible enemy attacks.

A huge uranium processing plant was being built secretly at Oak Ridge, Tennessee. The suggestion was made that the new laboratory be constructed nearby. Groves rejected this because he feared concentrating too much Manhattan activity in one location; that, too, could act to stimulate possible sabotage. Two more locations in California had been considered and rejected. The first, northeast of Los Angeles in San Bernadino, lacked suitable laboratory and living facilities; it was also thought by Groves to be too close to Los Angeles and therefore tempting to scientists to visit their friends and colleagues. The second location lay on the eastern side of the Sierra Nevada Mountains near Reno, Nevada. But there the heavy snows could inhibit work during the winter months.

Groves and Oppenheimer narrowed their search to several possibilities in New Mexico. The state contained large ranges of government-owned land and fulfilled the requirements of isolation and distance from both coastlines. A major rail center already existed in Albuquerque, and its airport had one of the newest and longest runways in the country. There were five possibilities: Gallup, Las Vegas, La Ventana, Jemez Springs, and a place suggested by Oppenheimer called Los Alamos. The first three locations were rejected by Groves because of poor space available for buildings. Only Jemez Springs and Los Alamos seemed good possibilities. Groves agreed to meet Oppenheimer in Albuquerque and drive north to check out each location.

Jemez Springs, he discovered, was a long, thin valley dramatically hemmed in on three sides by tall cliffs. Existing buildings left much to be desired, and even though the laboratory population was projected to be small, it was clear to Groves that perhaps 70 percent of the buildings needed would have to be constructed. Worse, an Army study ordered earlier by Groves revealed that the entire valley was subject to flooding. Everyone agreed that it was unacceptable.

Returning to their car, they drove north again toward Otowi and the Los Alamos Ranch School for Boys. From the reports compiled by the Army Corps of Engineers, Groves already knew that the location generally met all the preliminary criteria he and Oppenheimer had wanted. The area was isolated and it was set high on a mesa with limited access that could be controlled easily. A preliminary count of school buildings suggested that many of the required physical facilities already existed. It appeared from the survey report that, at least for the moment, there was an adequate supply of both water and electrical power.

The Army report only supported what Groves saw for himself. From the car he could see several large buildings and a number of small houses and ranch buildings. He was pleased to learn that some 46,000 acres of land of the estimated 54,000 acres required already lay in government hands. The remaining private land was primarily used for cattle grazing, and this fact suggested to Groves that the purchase price would be less than for crop land. The openness of the Plateau provided what seemed to be endless possibilities for expansion and safe spacing of buildings.

Groves and his companions drove as best they could on the dirt roads around the school complex. Everyone felt that Los Alamos was the only real choice. Groves returned to Albuquerque with orders churning in his mind. He left immediately for Washington to initiate his request for Los Alamos.

Connell was saddened but not surprised when the government sent word that his school—his mesa—would be needed for a wartime "installation." The sudden land surveys and mysterious visits by the Army were linked to the low-flying airplanes. The odd occurrences of the last few months finally made sense. The Corps of Engineers opened negotiations with Connell and the school's board of directors. Talk of leasing the land owned by the school to the government for the duration of the war ended when Connell received a personal letter from Secretary of War Henry Stimson indicating that the prosecution of the war necessitated the appropriation of his school and its accompanying land. Resistance was useless against such a move. Connell tacked the letter to the notice board outside his office for everyone to read.

At first the students and faculty were incredulous, and then saddened. There was no choice, however. Classes were accelerated and arrangements were made for most students to attend schools elsewhere. A year's work was compressed into six months. The school and classes concluded with the sound of machinery digging up the sleeping land; a new ambiance was replacing the old.

On March 22, 1943, Secretary Stimson requested approximately 45,000 acres of land under the jurisdiction of the U.S. Forest Service. Stimson made his request by saying that "there is a military necessity for the acquisition of approximately 54,000 acres of land for the establishment of a demolition range."[1] The remaining 9,000 acres would be private land. The government moved quickly. To avoid the endless legal complications that would be involved in normal land proceedings, War Department lawyers took all the privately-held acreage under Declaration of Taking procedures. This process removed title defects in acquiring privately-owned land.

Surprisingly, there was little resistance to the dealings; only two families on the mesa objected to the cash settlements offered by the government. There was a momentary snag when a Washington bureaucrat discovered that some of the designated land included Bandelier National Monument and an ancient and sacred Indian burial ground. After some debate, the war effort acquiesced to heritage: the boundaries of the new "demolition ground" were redrawn to sidestep Bandelier.

It was a good deal struck by Groves for the Manhattan Project: the final cost to the government barely reached $415,000. It was a sum that was to be one of the smaller Manhattan expenditures of the war. Everyone, however, including the Army engineer Groves, seriously miscalculated the scope of the new laboratory. Oppenheimer's original estimate was a laboratory with perhaps a hundred scientists and their families. In less than a year, this figure would jump into the thousands.

3. WORLD IN CHANGE

The Greek philosopher Democritus is credited with originating the concept of the "atom." The theory was a simple one by contemporary standards: that all matter was composed of tiny indivisible particles. He called these small pieces of matter *atomoi,* which literally means "uncuttable." Democritus was offering his new view of the physical world in Greece in the fourth century, B.C., just about the time that the Indians on the other side of the earth began to farm Pajarito Plateau.

It would take more than two thousand years for science to expand the theory of Democritus and to recognize that the nucleus of the atom contained the vast potentiality for providing a new source of energy. An Englishman, Sir James Chadwick, discovered the neutron in 1932: a small, electrically neutral particle within the nucleus of the atom. The discovery of the neutron was to be the key to unlocking the energy at the heart of all matter.

Chadwick's discovery was typical of the tremendous momentum gaining foot within physical studies. Since the turn of the century, scientists had been learning more and more about the nature of the atom and the forces that lay within it. Until the 1930's, however, most of the research in physics was conducted in Europe; only slowly had America emerged as a scientific power. In 1930, Ernest O. Lawrence constructed his first cyclotron, or atom smasher, at the Berkeley campus of the University of California, and in 1931, Robert J. Van de Graaff built his electrostatic generator for creating beams of subatomic particles.

By 1938, American scientists were only weeks behind their European colleagues in new discoveries. Phillip Abelson, a young Ph.D. candidate at Berkeley, was on the verge of a momentous discovery when his work was separately announced by Otto Hahn and Fritz Strassmann at the Kaiser Wilhelm Institute for Chemistry in Berlin. Working as a team, they had bombarded uranium with neutrons; the uranium was then analyzed and unexpectedly found to contain a radioactive barium isotope: Hahn and Strassmann had discovered nuclear fission. This discovery set in motion the forces of science and politics that culminated six years later in Hiroshima and Nagasaki.

For nearly a decade, during the 1930's, scientists in America and Europe had bombarded, or forced, beams of neutrons to strike various elements including uranium. Hahn and Strassmann had discovered that uranium could be "split" into two fragments through this bombardment process.* The splitting process was called *fission*. During fission, the uranium nucleus was divided into two parts with separate weights that put them approximately half-way between hydrogen, the lightest element, and uranium, the heaviest element known at the time. A chain reaction resulting from the fission of Uranium 235 atoms took the path shown in the illustration on page 22.

There were two important effects that occurred during fission: additional neutrons were suddenly freed from the original uranium nucleus, and large amounts of energy were also released. Each new neutron, if properly slowed down by a moderating material, could cause still another nucleus to split, releasing more neutrons and energy, and thereby repeating the process again and again. Repeated bombardment and splitting was called a *chain reaction*. Although it was not known at the time, one of the other side effects of a chain reaction under certain conditions was the production of another fissionable element called plutonium. This element would ultimately provide the guts of a second atomic bomb possibility.

News of the fission discovery spread quickly. Hahn im-

The form of uranium split by Hahn and Strassmann was later discovered to be the isotope Uranium 235.

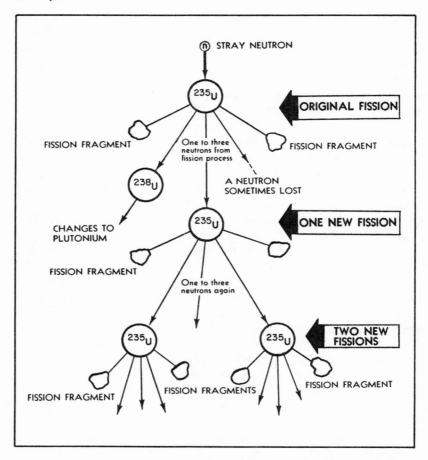

mediately recognized the importance of fission and contacted Lise Meitner, a colleague in Switzerland. Meitner and her nephew, Otto Frisch, quickly duplicated Hahn's experiment and verified that fission had taken place. Word was sent to Niels Bohr in England. News of the fission discovery was announced to scientists gathered at the Fifth Washington Conference on Theoretical Physics. Once in America, Bohr met with two other recent European refugees, Enrico Fermi and Isidor Rabi, to discuss the enormous implications of fission. This discovery suggested that a chain, or self-sustaining, reaction could take place. If the fission process released additional neutrons, then the splitting of, say, uranium nuclei, could in turn liberate vast amounts of energy. If controlled, the process could produce heat and power; if left uncontrolled, the same process could result in a powerful explosion.

American scientists began immediately to explore fission. At Columbia University, Fermi and Leo Szilard—another refugee from the Nazis—along with George Pegram, began to study the emission of neutrons during fission. Their experiments confirmed that neutrons were indeed emitted. They then turned their research to focus on the possibility that natural Uranium 238* might support a chain reaction. Another possibility was that a far more scarce form, or isotope, of uranium—Uranium 235—would do the same. Fermi set about testing the possibility of U 238 supporting a chain reaction; his colleagues in turn agreed to experiment with U 235.

Fermi found that a chain reaction would require two conditions: the neutrons must be slowed down to a rate at which they avoid being absorbed by the natural material around them, and they must be absorbed by uranium more often than by any other element. Slowing down neutrons, or "moderating" them, began another arm of major research.

In early 1939, Pegram wrote to Admiral Stanford Hooper about the work of Fermi and his colleague, John Dunning, at Columbia. It was possible, Pegram suggested, that uranium might be used to produce bombs tremendously more powerful than common explosives. While this was at the moment only a theoretical possibility, he urged that it be studied closely. Pegram indicated that Fermi would be in Washington soon and could meet with Hooper to discuss the matter. Pegram understated his case; the political events in Europe were alarming, and a new and powerful explosive weapon would certainly be of interest to any major power. Fermi met with naval representatives on March 17, but generated only mild interest.

Sensing the events in Europe, Leo Szilard also was concerned. He feared that the possibilities of fission bombs were already under study by the Nazis. Like Pegram, he believed that the new bomb remained largely a theoretical weapon, but one that nonetheless demanded immediate study. Szilard met with a fellow Hungarian, Eugene Wigner; together they decided that the matter needed to be brought to President Roosevelt's attention. But how? The two men decided that Albert Einstein had both the scientific authority and the international name to lead a campaign for government support of fission research.

Szilard was successful in persuading Einstein to draft a letter to Roosevelt. The eventual letter referred to work by Frédéric Joliot in France as well as that by Fermi and Szilard. Einstein suggested that a

---

*The number 238 refers to the atomic weight of the uranium atom; Uranium 238, or more simply, U 238, accounts for almost 99 percent of all uranium found in nature. U 235 is very scarce and accounts for approximately one percent.

chain reaction could under the right conditions generate enormous energy in the form of a bomb. Scientists needed the chance to see if these conditions could be produced. His letter also suggested that a liaison—an individual—be placed between Roosevelt's administration and the scientists undertaking fission research.

Szilard had also recruited the assistance of Alexander Sachs, an economist with access to Roosevelt. Einstein's letter was personally taken by Sachs to the White House on October 11, 1939. Sachs first met with General Edwin Watson, an assistant to Roosevelt, and several other military advisors to the President. After a preliminary review of fission research, Sachs was taken to meet with Roosevelt. Sachs repeated the implications of fission for a powerful bomb and emphasized the need for government support. The President listened carefully and said jokingly that he "didn't want the Nazis to blow them all up." Roosevelt told General Watson that action was needed.[1]

Roosevelt recently had appointed a Science Advisory Council that was headed by Karl T. Compton, president of the Massachusetts Institute of Technology. The President bypassed this council and established, instead, a new organization to be called the Advisory Committee on Uranium; as its head, Roosevelt appointed Lyman J. Briggs, director of the National Bureau of Standards.

Briggs quickly called a meeting of the new committee for October 31, 1939. Among those asked to attend were Sachs, Szilard, Wigner, and Edward Teller. There was already some doubt in scientific circles that a chain reaction could occur; the committee reviewed the research and felt it necessary to stress to Roosevelt that a chain reaction was indeed a possibility. If successful, the reaction could power submarines as well as provide the basis for bombs "with a destructiveness vastly greater than anything now known."[2]

An experiment by Dunning at Columbia in March 1940 again raised eyebrows. Dunning had obtained infinitesimal amounts of U 235 and was able to prove that this was the isotope that indeed fissioned with "slow" neutrons, or those that were slowed down by moderators. This discovery meant that a chain reaction was more likely to occur with U 235.

At an April 27 meeting of the Uranium Committee, Briggs, Sachs, and two others were joined by Pegram, Fermi, Szilard, and Wigner. The consensus of the meeting was that Fermi's moderator research needed to continue before they could recommend large-scale fission research. Szilard was disappointed. The months were slipping by without a strong and coordinated fission program.

In June 1940, Roosevelt created the National Defense Research Committee (NDRC), whose purpose was to coordinate scientific research behind defense needs. In Europe, Winston Churchill had assumed leadership of the British government and Germany had recently occupied Paris; Roosevelt felt the pressure growing and was not insensitive to what science might be able to contribute to a war effort. Roosevelt appointed Vannevar Bush of the Carnegie Institute as its chairman.

Bush quickly met with Compton, James B. Conant of Harvard, and Frank B. Jewett of the National Academy of Sciences, to discuss what role science might play in any future war. The recently created Advisory Committee on Uranium was transferred to Vannevar Bush and the NDRC. Briggs remained chairman, but the committee itself was reorganized to exclude foreign-born scientists. As it turned out, this was only temporary; within even a few months, much of the leadership of the fission movement would be in the hands of European scientists.

One of the NDRC's first concerns was to obtain a sufficient supply of uranium ore. Little of the material was used in the United States in 1940. After it was discovered that a New York company had some 1,200 tons of 65 percent uranium ore in storage, the committee shifted its attention to basic research. The committee believed that basic physics research must be given priority over chain reaction research. For the first effort, the committee approved $40,000, a sum that would be dwarfed within a few months.

Chain reaction research, however, fell behind schedule. The question remained: Were enough neutrons produced during fission to sustain a chain reaction? Fermi did not wait for NDRC support. He and his staff experimented and confirmed that graphite was a useful moderator for slowing down neutrons. Their work even suggested that placing uranium cubes uniformly throughout a stack of graphite blocks helped neutron absorption. The entire apparatus was called a pile. Briggs was impressed with Fermi's work. He placed an order for 8 tons of uranium oxide and 40 tons of graphite for Fermi to conduct an even larger pile experiment.

Vast quantities of uranium ore were needed, of course, in order to produce U 238 and U 235. These two substances differed slightly in weight, and only U 235 was fissionable* While the difference was infinitesimal by normal measures of weight, it was enormous by nuclear standards. One out of every one hundred uranium atoms was

---

*U 238 and U 235 differed not in atomic number but in weight; where the atomic number referred to the number of protons in the nucleus of the atom, the atomic weight referred to the combined total of protons and neutrons in the nucleus.

U 235, and methods were needed to separate out this scarce element from crude uranium ore. Large scale production of U 235 called not for a chemical process, but utilizing the slight difference in weight (mass) between it and U 238.

One possibility for separating uranium lay in using high-speed centrifuges. Such a centrifuge had been crudely built in 1939 at the University of Virginia. Small amounts of U 235 had been produced by using a gaseous mixture of uranium and fluorine gas known as uranium hexafluoride. The process produced only minute quantities of U 235.

Another possibility was gaseous diffusion, a process that involved forcing uranium hexafluoride through filters, or barriers. George Kistiakowsky, a professor of chemistry at Harvard, met in May 1940 with other scientists at the Carnegie Institute to review uranium production techniques. At the meeting he suggested that the diffusion process under development by the Bureau of Mines might be applicable to their needs. Kistiakowsky agreed to experiment with diffusion and to concentrate on developing barriers with sufficiently small holes to filter the uranium gas.

A still third possibility was liquid thermal diffusion. This process involved forcing gases to cool at differing rates and was similar to refinement techniques used by petroleum companies. The process seemed promising, and Briggs was prepared to allocate NDRC money to further experimentation.

The most promising method in 1940, however, was the electromagnetic separation process under development at Berkeley. Ernest Lawrence had begun work in the early 1930's on rudimentary particle accelerators, and within a few years had developed a series of larger, more sophisticated machines called cyclotrons. These machines used large magnets to keep charged particles accelerating in spirals within an electrical field. Lawrence and his staff were able to improve the process and by 1940 had been able to separate U 235 from U 238.

The electromagnetic process had great promise, but both Lawrence and the Uranium Committee realized that all four methods for uranium production must be pursued. Lawrence had full support for the electromagnetic process; both the centrifuge and thermal diffusion were distant bets at the moment.* The committee turned its attention to gaseous diffusion.

The sheer physical dimensions of diffusion were enormous. The process involved pumping—or pushing—uranium hexafluoride

---

After the war, some experts felt that the centrifuge method was the cheapest of the four.

through barriers in stage after stage. With each stage, the gas became increasingly purified and concentrated as U 235; the heavier U 238 could be collected at the bottom of the stages and the lighter U 235 could be collected at the top or end of the stages. Because natural uranium was 99 percent U 238, the number of stages necessary to separate out the U 235 were substantial. In effect, hundreds of miles of pipes and thousands of barriers would be needed.

But barrier design was the most perplexing problem. Many different substances were studied and discarded. The corrosion rate from the gas was dramatic and most materials were too brittle to withstand the extreme pressures. Nickel seemed the best bet with its high anticorrosive ability. The added complication lay in choosing a substance in which literally billions of holes less than one-thousandth of a millimeter could be made.

At the end of October 1942, Conant reviewed the separation problem with Bush. Of the four methods, the centrifuge seemed the weakest and Lawrence's electromagnetic method the most promising. No method seemed sure, however; both men felt saddled with inconclusive evidence and under great pressure to begin construction of plants to produce uranium.

There was a general feeling of paralysis across the fission movement. Despite the patient, thorough work conducted by Briggs, many American scientists were growing restless with the slow pace of research. In early 1941, the war in Europe had moved in favor of the Nazis. Intelligence sources reported that the Germans under Werner Heisenberg were working steadily on fission research. Lawrence, the recently appointed director of the University of California's Radiation Laboratory, spoke for many critics of the government's pace when he urged more money and more research.

A new and exciting discovery suddenly expanded the possibilities for fission bombs. Under Lawrence, Edwin McMillan had recently discovered Element 93, called neptunium, as the result of an experiment on fission fragments. Some months later, Glenn Seaborg, Joseph Kennedy, and Arthur Wahl produced an isotope of Element 93; there was, however, evidence of still another element. Within a few weeks, the group found proof of Element 94. This was particularly exciting because there was the possibility that this element would also prove fissionable like U 235. Within weeks, Emilio Segré confirmed that Element 94 was capable of fission. By the end of May 1941, Seaborg was able to tell Briggs that Element 94 was 1.7 times as likely as U 235 to undergo fission with slow neutrons.[3]

Seaborg's team at Chicago proposed to call the new element

plutonium, because it followed neptunium in the Periodic Table of Elements, and gave it the symbol Pu. The men discovered that the new element had two oxidation states; by treating one state with chemicals they could change it to the other state. This discovery had implications for processing the element in large quantities. By August, Seaborg was able to separate a visible amount of plutonium that was free from other matter. The sample, as well as its measurement, could be viewed only through a powerful microscope.

Infinitesimal as the sample was, its discovery was momentous. The new element meant another alternative to U 235 as fissionable bomb material. If, as Seaborg believed, plutonium was more likely to fission than U 235, then it was certainly worthy of producing in quantity in a large pile. Bush and Conant both felt that such a pile might mean quicker production of critical material. They immediately began to scout around for an industrial corporation with the skills to build such a plant. At that time, the only work on atomic piles was under Fermi at the University of Chicago's Metallurgical Laboratory.

Shortly after the plutonium discovery, Bush directed a review of the work of the Uranium Committee. Compton led the review, and along with Briggs, Pegram, and others, sought to assess the probabilities for a bomb and the possible date when such a weapon might be available according to research at the time. At a Harvard meeting on May 5, Compton heard from Kenneth Bainbridge, who had just returned from fission work in Britain. Bainbridge told the group that the British took the fission bomb very seriously; moreover, they believed the Germans were possibly ahead of the Allies in their work. The final report to Bush suggested that propulsion from fission for ships and submarines was three years away; a bomb made from U 235—possibly from plutonium—was perhaps three to five years away.[4]

Just a month later Roosevelt created still another governmental agency to coordinate scientific resources. The new organization was called the Office of Scientific Research and Development. OSRD was placed within the Office for Emergency Management directly under Roosevelt. In one bureaucratic brush stroke, many of the problems of financial support and coordination that plagued Bush were eliminated. The NDRC retained Conant as chairman, but the Uranium Committee went to Bush's OSRD and was then renamed the S-1 Committee.

The new S-1 was an enlargement of the Uranium Committee. Bush added Sam Allison and Edward Condon of Westinghouse to its board, as well as Lloyd Smith of Cornell University and Henry D. Smyth of Princeton University. New information lifted everyone's spirits: Segré offered more support for the belief that plutonium, or Pu 239 as it was called, was more promising than U 235 in sustaining a chain reaction. Word from the British revealed a new optimism: scientists in England saw the bomb developed in two years or less. Both American and British scientists now felt that a critical mass of five kilograms of U 235 would be enough for an explosive chain reaction.

On October 9, 1941, Bush met with Roosevelt and Vice-President Wallace. He explained the new developments and the new optimism. Roosevelt made several quick decisions: Bush must expedite present work but hold off on large-scale uranium plant production. Roosevelt stressed security; he restricted policy considerations to Bush, Conant, Wallace, Secretary of War Stimson, and Army Chief of Staff George Marshall. It was obvious to Bush that Roosevelt had larger plans for use of the bomb; he was also disappointed because now he would have to have direct approval from the President to make such a weapon.

On December 16, 1941, Vice-President Wallace called a high-level meeting to review the progress of fission research. At the meeting, Bush suggested that the Army assume control of the rapidly growing fission movement once full-scale work on uranium and plutonium production could begin. No one at the meeting disagreed, and the foundation for eventual Army control was laid. Later, in a conversation with Roosevelt, Bush received Presidential approval for the transfer, on the condition that full secrecy be maintained.

At Bush's direction, Compton organized a meeting in Schenectady, New York. Lawrence attended from Berkeley, as well as Robert Oppenheimer. The meeting suggested that perhaps as much as 100 kilograms would be needed for an effective U 235 bomb. Bush had asked that the conference send its report along to Roosevelt. The report dealt largely with technical issues. No less than two kilograms of U 235 could be used in a bomb and the need for "quick assembly" was crucial to success. The problem of predetonation was also raised: if a stray neutron entered the critical mass it might cause a chain reaction that would quickly fizzle and burn itself out.

The meeting also marked the entrance of Oppenheimer into America's bomb movement. Although his concerns at Berkeley were primarily theoretical, he nevertheless followed closely the developments in fission. Oppenheimer soon began to be a major spokesman for many theoretical considerations of the bomb.

Bush became all the more convinced of the need to explore—quickly—the various methods of obtaining large quantities of U 235. He sent Roosevelt a copy of the Schenectady proceedings, along with a letter indicating that he was forming an "engineering group" to look into plant designs; he closed with the notice that he would wait for

Presidential approval before committing any further resources to production. On January 19, 1942, Roosevelt returned the letter along with a handwritten notation: "V.B. OK—returned—I think you had best keep this in your own safe. FDR."[5]

Bush had permission. He began by taking on Edgar Murphee, a vice-president of the Standard Oil Company, and a man with sound experience in the construction of large chemical plants. Murphee began to look into the requirements of large-scale uranium production.

From the West Coast, Lawrence continued to urge Bush to put more money into plutonium research. Lawrence had already reported that Pu 239 had a spontaneous fission* rate no higher than that of U 235. This was encouraging, for now it seemed that Pu 239 might be an even quicker route to the bomb because it could theoretically be produced more quickly than U 235.

Bush used the emotional climate caused by Pearl Harbor to push for action. He informed Briggs, Compton, and Lawrence that Murphee was assuming responsibility for plant design. In addition, Bush created three new divisions within the S-1 Committee: Stan Urey would take charge of diffusion and centrifuge separation; Lawrence would continue to head the electromagnetic work at Berkeley; and Compton would continue with basic physics research.

A timetable was prepared by Compton: the possibility of a successful chain reaction must be determined by July 1, 1942, the actual test of a chain reaction concluded by January 1943, the first production of Pu 239 in a pile by January 1944, and a final bomb by January 1945.[6]

Bush also began to see the possibility of centralizing all fission research in one location. At the moment, he was spending a great deal of time simply traveling from project to project and holding meetings. Not all work could be centralized, of course, but key men and projects could be to some extent. There was a great deal of work going on in Chicago; Bush organized the Metallurgical Laboratory there at the end of January 1942. Richard Doan was made director, with Fermi, Allison, and Wigner in key roles.

Murphee was also able to make recommendations to Bush: at least one centrifuge and one gaseous-diffusion plant must be built to determine their effectiveness. As a corollary, the S-1 Committee must be sure to obtain sufficient quantities of uranium oxide, and a plutonium pile must also be built for mass production of Pu 239.

On May 23, 1942, Bush again called together Compton,

---

*Spontaneous fission refers to the characteristics of some nuclear materials, such as U 235 and Pu 239, to occasionally begin fissioning without being artificially bombarded from the outside by neutrons.

Conant, Briggs, Murphee, and Lawrence to review their programs and make recommendations for plant construction. By late afternoon, the group agreed that a centrifuge plant must be built by January 1944, with the cost approximated at $38 million; a gaseous-diffusion plant before the end of 1943, at a cost of $2 million; an electromagnetic plant to be completed by September 1943, at $12 million; one or more plutonium piles by January 1944, at $25 million; and one plant to produce heavy water by May 1943, at a cost of $2.8 million.

It had been an expensive day: nearly $80 million in expenditures had been recommended by the five men. No doubt Bush found the move ironic: barely a year before such a sum would have been unthinkable. The size of the recommended expenditures, however, underscored the rapid evolution of the fission movement. Bush was no longer comfortable with total responsibility for both design of a weapon and production of nuclear materials. The time had come to involve the Army directly.

4. THE NEW ALLIANCE

General Marshall appointed Brigadier General Wilhelm D. Styer as the principal liaison between the Army and the S-1 Committee early in 1942. Both Bush and Styer met with Marshall on June 10 to discuss the transfer of uranium and plutonium production to the Army. Marshall voiced serious concern that if all methods of production were attempted, the war effort might be jeopardized. He gave tentative approval to the pile and electromagnetic programs. Bush persisted; the realization of the bomb, he argued, rested fully on exploring all four methods.

The S-1 Committee had reviewed the work on all production alternatives and arrived at the recommendation that these four were the most likely to succeed in time for a bomb to be developed and used in the war. The impasse was resolved with Bush's adroit diplomacy. He persuaded Marshall to approve all four methods with the provision that priority be given to the methods requiring the least draw on critical materials and offering the highest promise of success. Marshall agreed, and Bush drafted a letter to President Roosevelt. He argued for both limited and all-out approaches to production, but with a subtle persuasiveness in favor of starting several production plants immediately. The letter concluded with a statement of the division of funds and responsibilities between OSRD and the Army: the Army would control 60 percent of the program in 1943, and all of it in 1944.[1]

Styer did not wait for Presidential approval; he immediately appointed Colonel James C. Marshall of the Army's Syracuse Engineering District to head the new fission program. Marshall was a West Point graduate with field experience in construction. He took command in late June and gave the new Army effort the simple designation DSM Project. Shortly thereafter he took the project to New York City and renamed it the Manhattan Engineering District.

Conant also proceeded to reorganize his resources. With Presidential approval on June 17, Conant recommended to Bush that a new S-1 Executive Committee be created to supervise all future OSRD work. Bush agreed, and Conant became chairman.* He and his committee were to oversee all technical research.

Soon, the new relationship between OSRD and the Army began to deteriorate. Colonel Marshall had taken command of an Army undertaking with a relatively low priority within Army ranks; indeed, because of the secrecy involved, only a handful of Army officials knew about it. His only good fortune was to have assigned to him Colonel Kenneth Nichols, who brought with him experience in both scientific research and engineering. Nichols, however, was just one man. The relationship between the Army and OSRD was young and vague: there was confusion, for example, over questions of who ran research relating to production. Most importantly, the "gentlemen's agreement" created between Bush and General Marshall, and approved by Roosevelt, did not generate a management process to resolve differences of opinion and authority. The lack of momentum within the Army threatened Bush's S-1 efforts.

Colonel Marshall lacked both rank and experience with bureaucratic procedures. A fine field officer, he now found himself hampered at every move by endless Army red tape. He was unable to obtain more than an A-3 war priority for procurement where nothing less than A-1, the highest possible priority, was needed.

Bush was dismayed. The fission project was again faltering, this time for lack of Army action and support. He went to General Marshall and obtained an endorsement to approach the War Production Board for a priority change. Quietly he sought to stir the Army to action

---

Conant's committee consisted of Briggs, Compton, Lawrence, Murphee, and Urey.

and to have Colonel Marshall replaced with someone more aggressive and sympathetic to the fission cause. The Army finally acted in September. Bush was notified that Marshall was being replaced with a new man. On September 17, Colonel Leslie R. Groves presented himself to Bush with the announcement that he had been appointed to direct the Army's Manhattan Project.

Groves was well qualified: he was a member of the Army's Quartermaster Corps with an impressive record. He had graduated fourth in his class from West Point and had been with the Engineering Corps in Hawaii, Nicaragua, and a half-dozen states. Most of his experience had been in large-scale construction. Only recently he had completed the new Pentagon Building in half the time estimated. He was a teetotaler with a fondness for chocolates. His experience was matched by his "by-the-book" personality and acerbic tongue. His personal style, in spite of his talents, had kept him back in Army ranks; for almost fifteen years he had remained a lieutenant. With World War II he had looked forward to a field command somewhere overseas. Instead, he was given the task of working with scientists to build several unorthodox industrial plants.

His disappointment, however, faded quickly and he sprang to action. Groves was promoted to Brigadier General on September 23, and within two days made a major decision to locate the gaseous diffusion plant at Oak Ridge, Tennessee—a decision that had stalled under Marshall for three months. He personally visited and badgered the War Production Board into granting the Manhattan District an A-1 priority rating.

Shortly after his promotion, Groves arranged a meeting on September 23 with General Marshall, Bush, Conant, and Generals Somervell and Styer. A Military Policy Committee was created to oversee the military's participation in the atomic bomb project. Groves insisted that the committee have only three members: Bush spoke for the scientists and acted as chairman. General Styer represented the Army, and Admiral William Purnell served as the representative for the Navy. Groves would report directly to the Policy Committee; to Bush's relief, Groves seemed in command.

In May 1942, Gregory Breit resigned as the coordinator of fast neutron research within the S-1 Committee. Compton quickly appointed Robert Oppenheimer to replace Breit. Oppenheimer's easy style and quick mind had impressed Compton and convinced him that Oppenheimer was the right man.

Oppenheimer had been a gifted child. After an easy graduation from Harvard in 1925, he had gone to study at the Cavendish Laboratories in Cambridge, England. It was there that quantum mechanics captured his attention and occupied his brilliant mind. From Cambridge he went to Germany to study under Max Born at Göttingen. Two years after his graduation from Harvard, Oppenheimer received his doctorate from the University of Göttingen, in 1927. After two more years in Leiden and Zürich, he grew homesick and returned to America to lecture at Harvard.

It was the beginning of a whirlwind that would lead him to both prominence and anguish. From Harvard he journeyed to the California Institute of Technology. During the long trip from east to west in a Cadillac convertible, he let his mind wander in and out of quantum physics and went off the road twice; the second time, he drove his car into the door of a local courthouse.

Oppenheimer had been offered positions at twelve universities, including two in Europe. With an inherited income of some $10,000 a year, Oppenheimer could be casual about the matter of where to work. He accepted positions at both Pasadena and Berkeley, hoping to shuttle from one to the other in spring and winter.

His pace quickened, and then his health failed with the onset of a light attack of tuberculosis. In 1929, Oppenheimer and his younger brother, Frank, went to spend the summer in New Mexico in the Sangre de Cristo Mountains outside Taos. Once there he fell in love with the New Mexico land and returned summer after summer. Now he had two loves: physics and New Mexico.

Oppenheimer recuperated and returned to California and social success. As a bachelor he, together with Ernest Lawrence, became the most sought-after young faculty men. Wives loved Oppenheimer at their parties for his charm and wit. Lawrence had begun his meteoric rise in scientific circles with the development of his early cyclotrons. During the 1930's, his work in physics helped to put American universities on a par with European schools. Oppenheimer, with a name better known in Europe, undertook long cerebral forays into studies of cosmic rays and collapsing suns. On the side, Lawrence dated pretty women and drank red wine; Oppenheimer dated little and read classical French literature and taught himself Sanskrit.

In 1939, Oppenheimer met an attractive young couple at Pasadena, Richard and Katherine Harrison. Richard Harrison was a British doctor interning at a local hospital, and his wife, Kitty, had recently obtained a degree in biology. The two had been married less than a year. Oppenheimer was immediately taken with Kitty, and in the course of the year the two fell in love. In October 1940, Kitty went to

Nevada to establish legal residence in order to obtain a divorce. On November 1, Robert Oppenheimer and Kitty were married. The two returned to Berkeley and submerged themselves in Robert's teaching and faculty life.

Oppenheimer became a favorite of his students. His quick mind and sensitivity to students won him a devoted following. Within a few years, his reputation was solidly entrenched: study under Oppenheimer was the highest recommendation for young theoretical physicists.

In May 1942, when Compton offered him Breit's position, Oppenheimer was involved in fission research as a theoretician. As a prerequisite to accepting Compton's offer, Oppenheimer asked for the assistance of someone more familiar with the current work in experimental physics. Compton agreed to assign him John Manley of the Metallurgical Laboratory. Manley, who was bright and informal, was engaged in cyclotron research at the University of Chicago.

Both men met in Chicago on June 6 and agreed that Oppenheimer would continue his theoretical work at Berkeley, while Manley would take over several projects in Chicago. In addition, Manley would direct the geographically scattered work of John Williams at Minnesota, Joseph McKibben at Wisconsin, and Norman Heydenburg at Carnegie Institute. Manley at first had reservations about working with Oppenheimer and expressed them to Compton. Several years earlier, Manley had heard Oppenheimer deliver a lecture and had been impressed with his thinking, but put off by his seeming aloofness and sense of detachment. Now he was being asked by Compton to work with Oppenheimer and to reassess his feelings. Manley quickly changed his mind, and the two became friends and a good working team.[2] Both men felt a responsibility to visit each project site and the next few months involved long hours of tedious travel. Manley found the work fascinating but exhausting.

In the midst of their travels, Oppenheimer saw the chance for planning and asked a small group of young physicists to join him and a close colleague, Robert Serber, at Berkeley for the summer. Oppenheimer arranged for several meeting rooms at the top of LeConte Hall; each window as covered by heavy steel mesh and the only door was fitted with a special key.

Edward Teller, Hans Bethe, and others came. The meeting in June produced the consensus that few theoretical gaps were left in neutron research, but that no one could be sure how much fissionable material was needed in the bomb. The best guess was that a ball of uranium some eight inches in diameter would be needed. Hard data were unavailable.

There was, however, a startling possibility revealed through the informal discussions. It was possible that a larger, more powerful thermonuclear bomb could be produced by igniting liquid deuterium with the heat from a simple uranium fission bomb. Such a bomb might even be able to ignite the atmosphere. This prospect seemed so important and ominous that Oppenheimer undertook a special journey to Compton's summer home to give him the news. Compton, however, like Bush, accepted the news by asking if such a weapon could be made in time for use in the war.

Bush and Compton were interested in delivery. Their interests at the moment centered on the production of uranium and plutonium. They were hardly interested in a weapon that depended critically on another weapon not yet developed. The plutonium-producing pile was now emerging as a quicker, less expensive way to delivery of fissionable material. Groves also felt that the new element might be the key to faster delivery of a usable weapon—if, of course, the pile could generate the necessary amounts. At the moment, need for atomic bomb production in the near future outweighed planning for a more distant "superbomb."

Groves and Conant met with representatives of the du Pont companies. Both men felt that du Pont would be the best choice to assist the Manhattan Project's major contractor, the Stone and Webster Company, in the design and operation of an atomic pile. The du Pont people took all the information Groves gave them and left for three days. On November 12, the du Pont officials agreed to undertake the task; they felt that there was a chance that they could delivery plutonium by sometime late in 1944.

Just two days later, on November 14, Bush and Conant learned that Fermi was nearly ready to put his latest uranium pile into operation at the University of Chicago. Fermi had taken possession of the university's unused squash courts in an abandoned stadium on the city's south side. On the same day, word arrived from Chadwick in England that plutonium appeared more likely to have spontaneous fission than uranium. This was contrary to earlier reports and meant that a bomb would simply need plutonium of a higher purity than was earlier thought.

On the night of December 1, 1942, Fermi and his staff completed the final assembly of a large pile of materials into a cube composed of nearly 400 tons of graphite, 6 tons of uranium metal, and 50 tons of enriched uranium oxide. The "enriched" oxide was simply uranium ore that had been processed to remove as many impurities as possible and then prepared in a metal form. Since early November, the team had been laboriously making graphite blocks, pressing the uranium into small pellets, and arranging both into a large square cubelike structure of alternating graphite blocks and uranium of differing concentrations. The entire pile had been enclosed in a large cloth balloon.

Tests conducted throughout the assembly revealed that the critical size of the pile would be reached sometime in the afternoon. Shortly after 4:00 P.M., the last layer of graphite and uranium bricks was put down and measurements made. Penetrating the pile were three sets of boron "control rods"; these long tubes of boron absorbed the neutrons and made a chain reaction impossible. Fermi was convinced that when the rods were removed a reaction would take place.

The next morning, December 2, at 8:30, Fermi, his staff, and invited guests gathered to witness the first test of a nuclear pile. Just at the north end of the courts was a small balcony about ten feet from the floor. Fermi, Walter Zinn, and Compton arranged themselves around a panel of instruments at the east end; the rest of the group stuck closely to the balcony. Several of the manual control rods were placed so that they could be dropped if the pile became too radioactively hot. A few staff members were prepared to flood the pile with a liquid solution of cadmium salts to retard the exchange of neutrons.

At 9:45 Fermi ordered the boron rods withdrawn. Fifteen minutes later the emergency rod was withdrawn. During the next hour the final rod was withdrawn by stages. Each time Fermi would accurately predict the neutron intensity as measured by dials and by a recording instrument. Shortly after 11:30, the pile's intensity continued to climb until suddenly the automatic safety rod dropped into the pile and shut it down. It wasn't until minutes later that the startled audience realized that the automatic safety level had been set too low.

Everyone broke for lunch and returned to witness the last rod being taken out, inch by inch. With all eyes craning to watch the dials, the drama continued until almost four o'clock. Fermi broke into a smile and announced that the reaction was self-sustaining. The time was 3:53 P.M. Science had just created a controlled chain reaction.

While Fermi and the others were celebrating the achievement with a bottle of Italian Chianti, Compton sent word to Conant at Harvard by telephone. He said, "You'll be interested to know that the Italian Navigator has landed in the New World." Asked how were the natives, Compton replied, "Very friendly."[3]

The chain reaction at Chicago gave Bush and Groves the hard proof they needed: now there was no doubt that under the right conditions an explosive chain reaction could be made to take place. Keen on the publicity import of the Chicago experiment, Groves prepared a report for his Military Policy Committee to approve and send on to Roosevelt.

The committee was now able to make a reasonable prediction of the bomb's future: there was a small chance of a weapon by June 1944; a somewhat better chance by January 1945; and a good chance by mid-1945.[4] The report also suggested that while it was unlikely that the Germans could have such a weapon by 1943 or 1944, it was possible that they might be as much as *six* months or a year ahead of the United States. This assessment had profound impact on the urgency of the Manhattan enterprise and upon the support that it was able to achieve.

The committee's request for massive funding was quickly approved by Roosevelt on December 28. Even before the President's approval, Groves had let contracts for nearly a half-billion dollars with American corporations. Bush approved the transfer to the Army of all nuclear-related contracts held by the Office of Scientific Research and Development. The empire held by Groves began to double and triple in size.

There were many factors for Groves to consider. With the President's approval in December, production of uranium and plutonium seemed assured in time. The task now was one of designing and fabricating a weapon. The nexus of bomb design at the end of 1942 was the Metallurgical Laboratory in Chicago. Groves and his Policy Committee now knew that the scattered fission work must be drawn together. Bush had already recommended that a joint OSRD-Army committee should be created to oversee the development of the bomb itself. Groves had strong feelings about compartmentalization of information; he had already learned of information leaks in Chicago, where news of the Fermi experiment had spread quickly. Shortly after the Berkeley conference, word leaked out from the Metallurgical Laboratory that work on a "superbomb" was underway. Bush tried to argue that an open exchange of information—at least at the top level—was in large part responsible for the successes achieved so far. Bush himself kept many of his project leaders in the dark about work elsewhere, but he and Groves both had the advantage of coordinating information and work from the top. A similar principle should be employed for bomb development.

With Oppenheimer's summer conference, the scientists themselves began to favor the establishment of a central laboratory. Oppenheimer soon led his colleagues in suggesting that the scattered research underway across the country could be better managed and its data better utilized if indeed all work could somehow be centralized.

Oppenheimer pressed this point with Groves shortly after his Army appointment. Oppenheimer cited the poor record of the Metallurgical Laboratory in security matters. The exchange of information, however, had been useful within the Lab itself. The concept of an open exchange of information with security could be maintained *if* the research teams were centralized and isolated.

Oppenheimer became convinced, as had some of his colleagues, that

> ... we needed a central laboratory devoted wholly to [work on the bomb] where people could talk freely with each other, where theoretical ideas and experimental findings could affect each other, where the waste and frustration and error of the many compartmentalized experimental studies could be eliminated, where we could come to grips with chemical, metallurgical, engineering and ordnance problems that had so far received no consideration.[5]

The idea of a central weapons laboratory fitted the pragmatic concern for security held by Groves. Oppenheimer's arguments for an isolated laboratory in which weapons work could be contained safely was realistic and—to Groves, at least—was a method of containing scientists within a military reservation.

The possibility of a special laboratory raised the question of a director for such an organization. For some time Groves had been impressed with Oppenheimer's quick mind and grasp of weapon requirements. Moreover, Oppenheimer had a distinguished record in theoretical physics. While he lacked management experience *per se*, Groves was most impressed with his ability to work with other scientists and to keep them within a constructive framework.

Groves sought suggestions on the matter. Compton initially favored Carl Anderson of Pasadena, a winner of the Nobel Prize in physics. Anderson, however, was unimpressed with the projected role of director. He rejected a tentative offer as beneath him. Lawrence was another choice, and one favored by Groves, but his vast electromagnetic plant at Berkeley needed his full attention during the next few years. There were several other possibilities that were, one by one, rejected by Groves.[6] At age thirty-eight, Oppenheimer became Groves' first choice for director of the new laboratory.

Oppenheimer had an unusual past. During an earlier period of his life—after his return from Europe—he had been peripherally involved in some of California's left-wing movements of the 1930's. He gave money to causes urged upon him by friends and was seen at

Spanish War Relief rallies. His wife, Kitty, had once been briefly married to a man who was a member of the American Communist Party, and to please him, had joined the party herself. Before his marriage, Oppenheimer dated Jean Tatlock, a young woman also involved in left-wing causes at Berkeley. For a time, she, too, claimed membership in the Communist Party. At the height of their relationship, several times they made and cancelled plans to be married. Even Frank Oppenheimer, Robert's brother, dabbled in Socialist causes, and at one point naively mailed in a coupon to become a Communist Party member.[7]

Despite these circumstances, General Groves never felt Oppenheimer's loyalty to be in question. As a matter of course, Groves had ordered security investigations for most major Manhattan Project figures. Oppenheimer's clearance was held up for months while junior investigators pried into his past. At all times, Oppenheimer provided information and even submitted to several long personal interviews with the project's security officers.

Groves was finally required, in July 1943, to issue the following order to the Security Office: "In accordance with my verbal directions of July 15, it is desired that clearance be issued for the employment of Julius Robert Oppenheimer without delay, irrespective of the information which you have concerning Mr. Oppenheimer. He is absolutely essential to the Project."[8]

Oppenheimer's lengthy investigation had been led by Lt. Colonel Boris T. Pash, a man who took a zealous interest in Oppenheimer's past. Early in his examination, Pash decided that Oppenheimer was a security risk, and despite the fact that no evidence of disloyalty had been found, wrote his superiors in Washington that Oppenheimer was still connected with the Communist Party.[9] Pash reported to Lieutenant Colonel John Landsdale, director of G-2 (Security) for the Manhattan Project. Landsdale was not unsympathetic to Pash's anti-Communism, but hardly found Oppenheimer the risk Pash made him out to be. For the moment, however, the matter simply ended with the dictum from Groves. Although Landsdale would continue with Manhattan security matters, Pash was removed from his investigations and placed in charge of a secret effort to gather intelligence on German atomic bomb activity.

The order by Groves lagged behind Oppenheimer's directorship of the new laboratory by nearly eight months. For all practical purposes, Robert Oppenheimer had been approved by Groves since the time that Los Alamos was chosen as the site for the development of new weapons. Groves was not unhappy with his choice; he prided himself on his ability to make quick decisions. Oppenheimer had been very useful during the fall and had demonstrated excellent rapport with his colleagues. His solid performance as a replacement for Breit, and as a leader during the Berkeley summer conference, convinced Groves that Oppenheimer was the right choice. Groves involved Oppenheimer in the key planning for the new laboratory and placed him in a leadership role in November 1942. His titular directorship became formal with a letter from Groves and Conant in February 1943.

The special weapons laboratory was about to become a reality. As it took shape, there was cemented an extraordinary alliance between science and the military that had begun a year before. With the transfer of all fission research to Groves and the Army, the military establishment now assumed leadership—and possession—of nuclear energy. Oppenheimer and his scientists now reported to Groves; Groves in turn reported to his Military Policy Committee, with Vannevar Bush the only civilian on the committee. Science and the military would become even closer partners during the next three years.

5. LABORATORY ON A HILL

Groves proposed to call the new laboratory Project Y. Within a few months of his choice of Oppenheimer and Los Alamos, Groves and his Manhattan Project had secured possession of the Boys' School with its student body still in session, and the federal government was seeking to obtain some 50,000 acres of land on Pajarito Plateau. Galvanizing his resources from Army ranks, Groves ordered a military post to be established in Los Alamos with 254 officers and men; 190 of these would be assigned to the new post as military police.

Groves and Oppenheimer were rushing from one end of the country to the other and left the Army's Albuquerque Engineering District* in charge of all the construction at Los Alamos. In March 1943, Groves was able to assemble a few of his Manhattan District staff in Albuquerque and take command from the Army Engineers.

Groves had taken Oppenheimer's advice that the new laboratory should be operated under the auspices of a major university.

---

*The Albuquerque Engineering District was part of the Army Corps of Engineers and not part of the Manhattan Engineering District.

Both men believed that the use of a contractor afforded the new project an administrative expertise and a measure of security disguise. It would also make it easier for Oppenheimer to recruit his fellow scientists. The contract was given to Oppenheimer's former employer, the University of California. The university was to act in the role of prime contractor and be responsible for the procurement and the final operations of Project Y. A letter of intent signed on January 1, 1943, and formalized as an agreement on April 20, gave the University the responsibility—with "the utmost dispatch"—for the conduct of "certain studies and experimental investigations at a Laboratory located at a site which has or will be informally made known to the Contractor."[1]

An initial laboratory plan, drawn by Oppenheimer, Manley, and Edward McMillan, set the Los Alamos scientific population at about one hundred. Although Groves had originally discouraged families, Oppenheimer argued that staff would be hard to recruit if families could not be brought along. Housing needs were projected on the basis of residence of a hundred or so men and their families, plus some technical staff, and a few others.

The original "technical complex" was to consist of an administrative building and various laboratories and shops. The Army chose to keep Fuller Lodge, the largest building of the school. It was a two-story structure built of logs and would serve as a dining room and hotel for visitors. Just south of the Lodge was a small pool of water called Ashley Pond and a stone building that had served as an icehouse. Across a dirt road was the Main Technical Building, or T Building. It served as headquarters for Oppenheimer and his administrative staff and the Theoretical Physics group, and held a library with a vault for classified documents. Just behind T, and connected by a covered walkway, was Building U for the chemistry and physics laboratories. On either end of U Building were separate laboratories for the Van de Graaff and Cockcroft-Walton accelerators. V Building contained the shops. Almost on the edge of Los Alamos Canyon were Buildings Y and X for the cryogenics laboratory and the cyclotron.

Within a year, the Laboratory was forced to expand to accommodate the ever-increasing numbers of scientific and technical staff. Buildings to house the rapidly growing Theoretical Division were built across the road from T. The buildings were connected by second-story covered walkways, under which traffic could pass. Just behind the shops, the Laboratory constructed its first plant for the purification of plutonium. Warehouses, meeting rooms, and more administrative offices soon covered the congested mesa. By the end of 1943, very little was left of the original Boys' School. The entire Technical Area was encompassed by barbed-wire fences with several gatehouses and the area was patrolled day and night by guards on horseback.

Oppenheimer's office was first located in T Building, on the second floor. The new director chose a small corner office with a second adjacent room for his personal secretary. Priscilla Green had made the move from Berkeley with Oppenheimer in March 1943. At Berkeley, she had been a secretary to Ernest Lawrence, and first worked with Oppenheimer during the summer conferences the year before. With the organization of Los Alamos, she suggested to Oppenheimer that he needed a secretary and persuaded him to let her come to New Mexico.

Once on the Hill, her life became frantic. As secretary to Oppenheimer, Ms. Green wrote and answered letters, attended all meetings in the director's office as recorder, and for a while ran the mail room and the Laboratory's switchboard. She also made several harrowing drives with Oppenheimer on the ancient road from Los Alamos, taking dictation. The new director seemed particularly interested in recording conversations with Robert Underhill, the University of California's representative, and with General Groves. Despite protestations from Groves, Oppenheimer insisted that his secretary take notes on all conversations conducted over the telephone.

Their offices in Los Alamos quickly grew too small with the great influx of contractors and arriving scientists. In early 1944, they moved across the dirt road to A Building and occupied a much larger suite of rooms on the second floor. Oppenheimer settled in a room with one wall of windows that looked out into the New Mexico mountains. Large blackboards covered several walls, and a large conference table took up the far end of the room. Participants in Oppenheimer's many meetings sat in director's chairs made of wood and canvas. However out of place they seemed in Oppenheimer's office, they were considered luxury items by scientists who quickly learned what Army Supply considered attractive furniture.

During construction of the Tech Area, as it became known, it was suddenly realized that housing for expected staff was almost nonexistent. Both John Manley, who had been assigned by Compton to help plan the Laboratory, and Oppenheimer had seriously underestimated the housing needed. Some personnel were taken off Tech Area construction and set to building houses. Very little was ready at Los Alamos when Oppenheimer and a few hearty staff members arrived in Santa Fe on March 15. For the next few months, staff members and their families had to be located at scattered dude ranches in the area. Groves had ordered that no Laboratory staff stay in Santa Fe for security reasons; as a concession, however, to the disarray on the Plateau, he allowed a small contingent to set up an office at 109 East Palace Street, near the city's ancient downtown area. The purpose of the office was to answer delivery queries and to greet confused scientists. The Santa Fe office was turned over to a congenial and patient woman named Dorothy McKibben to manage.

Once arrived in Santa Fe, everyone had to make the tortuous 35-mile trip to Los Alamos by car or truck. The roads to the guest ranches were even more frightful, and living conditions were rustic. Families often had to crowd into small rooms and share kitchens and bathrooms. Transportation was as crude as the roads. The road to Los Alamos was winding, filled with difficult switchbacks, and layered with sharp rocks. There were very few official cars and trucks, and what was available was often old and prone to collapsing at inconvenient moments.

Until the middle of April 1943, only brief telephone conversations over an ancient Forest Service line were possible between Los Alamos and the Santa Fe office. Brief requests or instructions could be shouted; more complex conversations required a 70-mile round trip by car or truck. It was cold until late April, and few buildings were heated. Not even the commissary was open, and box lunches had to be imported from Santa Fe every day. The working day tended to be short and irregular for the first few months in 1943.

Most of the technical buildings were rather flat, prefabricated structures in what was called modified mobilization style. Each one contained drab exteriors of clapboard sidings with simple pitched roofs of asphalt or wood shingles. Only a few of the Tech Area buildings had air conditioning and dustproof construction. Most had acoustical tile ceilings. Buildings for the Army personnel were even less attractive.

Most new families to the Hill suffered some form of cultural shock. Everyone was told that he or she would be sent to New Mexico, and a few were told that the new laboratory was in the Jemez Mountains. For the non-Spanish speaker, the word "Jemez" sounded like "Hamos" and no map of New Mexico consulted revealed such mountains. As they made the initial drive through the beautiful New Mexico scenery, there quickly came the appalling realization that the squalid military town they were seeing was actually Los Alamos. They were even more appalled when they were shuttled off to live in rustic ranches, leaving their furnishings en route or in storage for the moment. What little housing there was bore the ubiquitous clapboard siding. The ambiance of Los Alamos had become military.

Each new contingent of arriving family and contractor suffered the road system. The Boys' School had been concerned with protecting the natural state of the mesa and had left road building to a minimum. Groves finally had to order that the road from Santa Fe be widened and paved to bear the heavy traffic of individuals, trucks, and deliveries that were arriving every day. Roads laid in winter, when the ground was frozen, melted with spring and turned to mud. There were very few paved roads on the mesa, with most housing arranged here and there and connected by dirt trails. When it rained, the plateau settled into a staggering mire that slowed all movement to a crawl.

The original assessment of Los Alamos by Oppenheimer and Groves seemed fated to be wrong at every turn. No sooner had men and machines begun to arrive than the water supply ran short. Teams were sent to tap water in Los Alamos Canyon, and later, the canyons of Pajarito and Guaje. The reservoir on the Hill grew algae. Winter weather froze and shattered pipelines. One huge water tank, then another, was finally built.

Oppenheimer and his advance guard arrived to find that Los Alamos now needed to provide the arriving hordes with sewerage systems, schools, stores, laundry, post office, telephones, garbage disposal, medical services, and some police protection. As soon as scientists arrived and began work, they began their building requests. Everybody needed something. The original contractor completed 54 percent of his contract in two months and the remaining 46 percent in another year. Everyone's request for expanded or new facilities delayed progress or shifted construction teams to another project or side of the hill. Oppenheimer's solution was to spend more money and hire more contractors.

The Los Alamos Laboratory formally opened on April 15, 1943. Groves was present to assist in the orientation with a limp handshake. As one scientist recalled, it was a hell of a mess: few buildings were completed, and there was mud everywhere. Oppenheimer could not afford to be discouraged. Instead, he arranged for a series of lectures to be given to new staff on the status of fission research.

Some men, like Manley and Serber, already knew much of what was going on. Most, however, were ignorant of the work as a whole and came from specific projects. Most, in fact, were young—the average age was in the twenties—and many were there straight from graduate schools. The lectures were to stem from a conference held in Berkeley the month before. Oppenheimer, Serber, and others had met with Richard Tolman of the National Defense Research Committee to review the status of all uranium and plutonium research. The result was a report prepared for Groves. The same report would in essence form the basis for the new-staff lectures. Robert Serber, a shy and hesitant speaker, was chosen by Oppenheimer to deliver the overview.

What was not entirely clear to Oppenheimer, however, was how to take the status of fission research and prepare a statement of work for Los Alamos. To assist Oppenheimer, Groves created a special Review Committee. He felt that a blue-ribbon panel was needed to help set priorities. Bush and Conant agreed, and Groves finally appointed W. K. Lewis, chairman of MIT; E. L. Rose, director of research for Jones and Lamson Machine Company; J. H. Van Vleck and E. B. Wilson of Harvard University; and Richard C. Tolman to represent the NDRC. Tolman was Bush's assistant and was to remain keenly involved in the affairs of Los Alamos throughout the war.

Oppenheimer and Tolman prepared a list of key research questions for the committee to review. These questions included critical mass, the conditions that affect nuclear reactions and detonation, the chemistry and metallurgy of uranium and plutonium, and the possibility of a thermonuclear, or hydrogen, bomb. The hydrogen, or Superbomb, required a fission bomb as a base to produce the incredibly high temperature necessary to ignite deuterium. Equally important, they said, were the production and delivery schedules of uranium and plutonium. Even at this stage there was a need to know what logistical requirements the military had for the bomb itself.[2]

Both the Review Committee and the new staff were briefed on the theoretical base for nuclear weapons and the extent of the technology known for putting a bomb together at the time. The lectures began on April 15, the Laboratory's official opening day.

The weather was cool with traces of winter as the scientists arrived at the recently completed Gamma Building. Like most other new buildings in Los Alamos, the roof was green and the exterior walls were of clapboard siding. Gamma had been built by the Laboratory as an auditorium for meetings. At one end was a small stage and wood and canvas chairs provided seating.

Oppenheimer introduced Serber, who began with fundamentals. It was known, he said, that U 235 contained 92 protons and 142 neutrons. When the nucleus absorbed an additional neutron, it became unstable and usually divided in half. This was the essence of the fission process. These two fragments together were then something less than the mass of the original uranium nucleus plus the additional neutron. Most of this difference was converted into kinetic, or explosive, energy.

No one knew, however, how many neutrons were emitted

from the nuclei for every U 235 nucleus split. Some U 235 nuclei might absorb these neutrons and therefore undergo fission, in turn to produce still more neutrons. The process, called a chain reaction, would proceed very quickly as long as one neutron from each fission caused another fission. Oppenheimer and his team at Berkeley had theorized that the energy release from one kilogram (about 2.2 pounds) of U 235 would equal the energy release from the detonation of 17,000 tons of TNT.

Given this combination of fact and speculation, the problem was to devise a bomb that received its explosive energy from the fission of U 235. It was also possible for a similar process to occur from the fission of Pu 239. Neither uranium nor plutonium was immediately available in sufficient quantity to make a bomb, or even, in 1943, to conduct many of the experiments that would provide needed information to design a bomb. Two large plants for production of U 235 and plutonium were only under construction at Oak Ridge, Tennessee, and Hanford, Washington.

The obvious task facing Los Alamos was to discover the process and procedure for making the desired liberation of energy take place *efficiently* and at *exactly* the right time. No one knew at the moment how much fissionable material had to be used to cause a chain reaction, but it was certain that nothing would happen if too little material was used. The fission process differed from that of conventional explosives in that a small amount of TNT, for example, burned as readily as a large amount. Fission chains depended upon neutrons remaining within the mass until they encountered other fissionable nuclei. The larger the mass, the easier it was for neutrons to find other nuclei to strike and cause additional fissioning. The best guess at the moment, Serber told his fascinated audience, was some 15 kilograms of U 235 and 5 kilograms of plutonium.

There were additional unknowns, Serber went on. No one had measured the exact number of neutrons emitted in fission and neither did they know the "cross sections" for nuclear materials. Knowledge of cross sections was critical to the scientists because it was an index of the likelihood that a given amount of uranium or plutonium could support a nuclear reaction.

"Cross sections" could be compared to the carnival game in which a person throws a ball at a series of milk bottles. If the surface area of the bottle is, say, one square foot, then the "effective" area—or cross section—that an individual can hit with a ball is one square foot. Cross sections become complicated when one varies the surface of the bottle or changes the object being thrown from a baseball to a Ping-Pong ball. And what happens when the individual throws a fast or slow ball?

In the case of fission, the bottle was the uranium or

plutonium, and the ball was the neutron. It was necessary to learn whether a free neutron would be likely to cause fission or merely pass through a sphere of pure U 235 metal. It was not important to learn how large, *per se,* the U 235 nucleus was, but rather how large a target it statistically presented for a neutron of a given velocity to cause a fission reaction. It *was* necessary to know the fission cross sections of U 235 for neutrons of different speeds before a weapon could be designed. Moreover, the uranium would not be pure U 235, and therefore the men needed to learn cross sections that accounted for various impurities—especially U 238.

It was clear to everyone that a major effort at the Laboratory would be to understand the concept of cross sections. The particle accelerators known as atom smashers could be used in these experiments because they could produce indirect beams of neutrons to bombard samples of potential bomb materials.

As important as cross sections was the concept of "critical mass," or the amount of fissionable material that was just sufficient to sustain a chain reaction, Serber continued. It would be necessary to surround this material with a covering, or envelope, or another material that could bounce back escaping neutrons into the uranium or plutonium core. The purpose of the reflecting shield was to develop an "economy of neutrons" by returning escaping neutrons back into the critical mass, where they were needed for a reaction. This shield was nicknamed a tamper, and it served another purpose: as the fissionable material expanded during the explosion, it quickly became less dense, and at the same time the surface area increased. The effect of the expansion was to increase the surface area and facilitate neutron escape. The tamper would serve to slow the expansion and to allow more energy to be generated before the reaction could be quenched.

The scientists were also told that the cross sections for U 235 were greater for slow neutrons—those slowed by moderators—than for fast ones. A nuclear explosion nevertheless required fast neutrons. An explosive chain reaction required one fission to follow another as rapidly as possible. This fact was important because the neutrons produced during fission were naturally fast, but most importantly, because slow neutrons could not liberate enough potential energy in time. In addition, a bomb designed for use in the field must be as light as possible, and cannot depend upon neutrons slowed down by moderators like graphite.

Still, a bomb must have a source of neutrons. Oppenheimer and the others did not think it advisable to depend solely on background neutrons to start a chain reaction. These neutrons were those that existed freely in materials at all times, or came from cosmic rays reaching the earth. There needed to be a constant, predictable source of neutrons that could be released at the right time. A bomb would need an internal source that would release hundreds of thousands of neutrons in a single burst at precisely the right instant. The men gave it the name initiator.

The speed of nuclear processes was astounding. Scientists at Los Alamos were talking about amounts of time measured in millionths of a second. The moment of assembly—when all forces in a bomb came together—was critical to success, and to the human mind at least, instantaneous. The problems were confounding: not only did Los Alamos have to have pure uranium or plutonium, but the scientists had to bring two subcritical masses together as quickly as possible and under precisely the right conditions. Dynamite was capable of exploding whenever its cap or ignitor was set off; a critical mass of fissionable material was not only capable of sustaining a chain reaction, it was also incapable of *not* doing so. Therefore, a bomb would work only when its nuclear core was assembled for the first time. Unlike a conventional explosive, which could be set off only with a percussion cap or fuze, a critical mass could be set off by any free neutron supplied by cosmic rays, neutrons from impurities, or a dozen other sources within the bomb.

Again, the scientists were faced with the incredible speed of nuclear reactions. Rapid assembly was crucial: as the uranium (or plutonium) core passed from a safe or subcritical stage to an explosive or supercritical stage, it would inevitably pass through configurations that were barely critical. If the process lingered too long in any of the intermediate configurations, the explosion would fizzle. A propellant would be needed to force the assembly to take place quickly. Another problem: What kind and what amount of gunpowder would be necessary?

The lectures ended for the day. Serber continued the next day with a discussion of possible ways to construct a bomb with uranium or plutonium. He outlined three general methods of assembly.

The first was the so-called "gun" method in which one subcritical mass of fissionable material would be fired into another subcritical mass. When projectile met target, there would be a supercritical, or explosive, reaction. Oppenheimer had Tolman's earlier report mimeographed and handed out: it included Tolman's original sketch of the gun's critical mass:

52/ Alpha

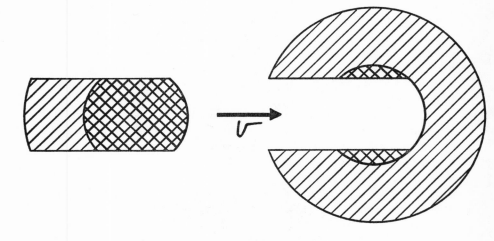

The gun method would have to employ a large bomb casing with an explosive charge and subcritical mass on one end and the remaining mass resting in a "seating" mechanism at the other end.

Serber suggested that the critical mass could be shaped into a sphere with a hole bored into the center. From the opposite end of the gun another portion of the critical mass, shaped in the form of the core, could be fired into the sphere. Timing was critical: if the small piece was fired too slowly, the chain reaction would fizzle.

The second method was a self-assembling, or autocatalytic, process which operated by the compression or expulsion of neutron clusters imbedded into the U 235 or Pu 239. Both Oppenheimer and Serber expressed some skepticism, however. Preliminary calculations suggested that this method, as it was then understood, would require large amounts of fissionable material and would give a relatively low explosive force. Tolman's sketch (p. 53) looked like the gun weapon.

The third method was by "implosion," in which a slightly subcritical mass of fissionable material would be surrounded by high explosives. When the explosives were detonated and forced inward, the subcritical mass would be forced into a supercritical stage and explode. Implosion was more difficult to achieve, Serber said. In the audience sat Seth Neddermeyer, a thin, frail man, who was quiet throughout most of the lecture. The concept of implosion captured his imagination and he tossed the idea back and forth in his mind. Neddermeyer's experience epitomized the purpose of open scientific exchange. Very shortly, he

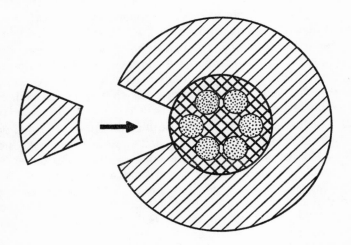

would develop the idea of implosion into a serious alternative and become the spokesman for it within the Laboratory.

While it was hoped that the gun method would work with both U 235 and Pu 239, no one could be sure. In fact, at the time of the April lectures, no one knew with certainty that plutonium emitted neutrons during fission.

The lectures raised again the possibility of developing a thermonuclear, or superbomb. Edward Teller rose from his chair to excitedly suggest to the audience that a fission bomb could be the basis for generating the necessary—and astronomically high—temperatures for igniting a substance like deuterium. Both the uranium and plutonium bombs utilized heavy nuclei as the basis for fissioning. Lighter elements would be used in a thermonuclear bomb.*

Deuterium was an isotope of hydrogen (the lightest element) that was twice as heavy as the hydrogen nucleus. Teller suggested that deuterium in a superbomb had several obvious advantages: it had the lowest ignition temperature (about 400 million degrees) and once ignited it was five times as explosive, or energy-producing, as U 235. The bomb had destructive effects vastly greater than those of the fission bomb. Teller argued that it could not be ignored by Los Alamos.

The superbomb discussion also surfaced a concern for reac-

---

*Among the lighter elements, the typical energy-producing or exoergic reaction is the building up of heavier nuclei from lighter ones. The energy that is liberated in the process goes into kinetic energy and radiation. Such a reaction in a mass of deuterium will spread under conditions in which the atoms are heated to very high temperatures; hence the process is called thermonuclear.

tions affecting the Earth and its atmosphere. Even as early as 1942, scientists had discussed the possibility that an atomic bomb might ignite a thermonuclear reaction in the light elements of the earth's atmosphere or its ground surface. A reaction with nitrogen nuclei in the atmosphere was the obvious possibility. Mathematical calculations, however, suggested that no matter what temperature a bomb produced, energy loss would be greater than energy production. The world seemed safe.

It seemed that "science and common sense" ruled out the possibility of a cataclysm. A report issued some years later suggested the Coulomb forces of the nucleus prevented such a catastrophe. Only a thermonuclear reaction could cause such an ignition, and, with a comparison to "stellar explosions, novae, and supernovae," it was decided that these forces were gravitational rather than nuclear in origin.[3] At the time, however, no one really knew what would happen, but everyone chose the mathematical projections.

Teller, like several of his friends at Los Alamos, had come to America during the 1930's to avoid the growing persecution of Jews in Germany and Europe. He had been invited by Oppenheimer to come from the University of Chicago to the new Laboratory to be a part of the Theoretical Division. From the very beginning, Teller's real interest lay in the possibilities for a hydrogen bomb. This interest kept him from any major work on the development of fission weapons, and caused some ill-will among his colleagues. His persistence, however, earned him a major role in the new weapon's development after the war.

Teller remained excited about the "superbomb," despite the many unknown factors to consider. He even persuaded Oppenheimer to assign him additional manpower to further theorization. Both men knew that the cost of developing such a weapon would be enormous; Teller wryly coined the term "megabucks" when referring to its cost. For the moment at least, Teller managed to introduce the term "superbomb" into the Los Alamos vocabulary.

When Serber's formal lectures ended, Oppenheimer opened discussion by all present. Everyone was astounded and pleased at the lack of compartmentalization. The intellectual atmosphere was heady and stimulating. The lectures and subsequent discussions also helped the Review Committee to formulate a framework of research and development for Oppenheimer to implement. Fundamentally, their work would be governed by two simple factors: research schedules would be dominated by uranium and plutonium production schedules, and the end product would be a bomb. Both U 235 and Pu 239 would not be available for perhaps two years; the design of the bomb would have to proceed immediately.[4]

The Laboratory's theoretical program had to focus on understanding and analyzing what went on in a nuclear reaction and what conditions had to be met to produce one. Which of three fissionable materials—U 235, Pu 239, and a new uranium hydride—appeared to be the most likely to produce the base for a nuclear weapon? Urgent research questions involved comprehending cross sections for all materials and for finding which materials could be used most effectively in a tamper. And how likely was it that a superbomb could be produced from a fission bomb?

For the experimental physicists, the committee recommended extensive detailed and integral experiments. Detailed, or differential, experiments involved determining the average number of neutrons per fission in both U 235 and Pu 239. These experiments would tell Groves and the Military Policy Committee about the wisdom of producing plutonium in quantity. The so-called integral experiments needed to focus on the process, or procedures, for bringing all components of a weapon together to produce an explosion.

There was certainly a full load for the chemical and metallurgical teams. They would not only need to learn more about the metals themselves, but also find new ways of refining and handling them. No one had ever handled large amounts of either U 235 or Pu 239; how did one cast or shape it? What about the initiator, or neutron source?

More laborious, perhaps, was the work assigned to the ordnance people. These men needed to learn how to bring subcritical masses together quickly enough to prevent predetonation. The critical mass, tamper, and explosive—plus the firing mechanisms—need to fit conveniently into an airborne bomb for use by the military. There were really two bombs to develop: gun and implosion. Two weapons needed to be designed, tested, and delivered to the Army. There was at least some experience with gun weapons, but implosion was a virginal field.

The Review Committee sent its report to Groves on May 10. Their recommendation, accepted and formalized by Groves, irreversibly enlarged the role of Los Alamos. Not only would the Laboratory have to design and fabricate bombs, but now it would also have to undertake extensive chemical and metallurgical work. It was clear to the committee that Los Alamos would have to undertake the final purification of plutonium and uranium. Even worse, work on the actual specification of the bomb would have to proceed immediately without the benefit of experimental data on U 235 and Pu 239. It would be at least two years before the rest of the Manhattan Project could deliver any real quantity of uranium and plutonium. Their work would begin, as one scientist said, on a very tentative note.

Within only a few week of the Laboratory's official opening in April, Oppenheimer found himself at odds with his original concept of "perhaps a hundred scientists." It was suddenly clear that there was not enough manpower available for the tasks ahead. Almost as bad, the mesa seemed very small. Men were needed, and would arrive with their families. Where would they live? Where would the huge plant for purifying plutonium be built? Had this diverse role for the Laboratory been seen earlier—even as recently as five months before—Los Alamos and Pajarito might not have been selected as a site for Project Y.[5] Soon enough, 55,000 acres would seem too small and a hundred men a painful joke.

6. THE ACTORS

Julius Robert Oppenheimer* was the common element in Los Alamos; he was the master weaver producing the unique fabric. Oppenheimer personally recruited most of his senior staff and came to know most of the others by their first names. He allowed himself to be nicknamed Oppie as a measure of affection and, in varying degrees, provided inspiration, punishment and reward, management, and a general emotional massage. He drove himself mercilessly in order to make Los Alamos work. His weight fell from 130 pounds in 1943 to barely 115 at the war's end. He smoked constantly, consuming one cigarette after another or smoked his pipe.

    Even Oppenheimer fell victim to the Los Alamos climate. He, like others from big cities, was forced to shed the more formal "university" attire in favor of comfortable clothing. His tweedy coats and thin ties seemed to exaggerate his slenderness as the years wore on. His porkpie hat became emblematic of the Laboratory and one jocose

---

*Oppenheimer always preferred to use the initial "J" instead of the name Julius. His birth certificate, however, registers him formally as Julius Robert Oppenheimer.

memorandum to the Maintenance Group requested a single nail for Oppenheimer's office and the "Director's hat." Oppenheimer later reflected that his best times at Los Alamos were those spent touring the Laboratory's scattered offices and sites; he seemed to thrive on the exchange between himself and his colleagues. Almost always, wherever he went he was followed by a small crowd of eager men. Their dialogue was usually fast and piercing. Observers often remarked how Oppie and his "children" looked like a mother hen and her chicks.

    Oppenheimer was marvelous at mediating differences between his powerful and eccentric staff. He was also sensitive to the moment when his presence or inquiry would be useful in unlocking an intellectual or developmental obstacle. Oppenheimer was both generous and irritating, for he could deftly lead his colleagues or chafe them with his intellectual superiority. Although it was rarely intended, there were moments when his arrogance offended some—usually those whose minds were less quick than his. At times he could be piqued at slowness and become condescending, and occasionally he would verbally strike at an individual with little cause, leaving nearby staff members stunned and embarrassed. On other occasions, Oppenheimer could hesitate in making decisions, especially where old friends were involved. In the most severe case, implosion research nearly reached an impasse before he acted to replace an old colleague and appoint new leadership over the critical work. And few men can remember Oppenheimer's giving a direct compliment.

    But undoubtedly his best moments were those in leading meetings among his colleagues, or settling differences, and acting as a buffer between General Groves and the military and the civilian staff. Oppenheimer courted Groves, flattered him, acted as the punctilious Director, and usually won his way over the tough General. In meetings among his staff, Oppenheimer set the agenda, listened carefully, and gave leave to individuals to ramble, but always skillfully pulled together a dozen alternatives into a cohesive direction.

    In the laboratories, his informal visits allowed him to watch and listen quietly from the back of the room and suddenly offer a new interpretation or solution. Although his presence intimidated some, most appreciated Oppenheimer's ability to see quickly many dimensions of a problem. His mind was not restricted to theoretical physics—his specialty—but was able to digest information and problems that embraced all concerns of the Laboratory, from metallurgy to explosives. On one occasion, Oppenheimer walked into a small room where physicists had spent a long day discussing a problem that filled a large blackboard with elaborate calculations. Oppenheimer observed for a moment, then walked to the blackboard and corrected a figure and walked out. Instances like this helped to promote an "Oppie myth" that the Director was able to exploit in his management of his diverse staff.

Oppenheimer was forced to draw on many resources to keep the Laboratory working. Since he was the Director, there were few persistent problems that did not eventually find their way to his office. He was constantly asked to handle housing inequities, explain why one division had better facilities than another, and assume ultimate responsibility for delayed supplies, faltering construction deadlines, and inadequate food and personal items at the PX. Oppenheimer usually listened attentively to complaints. Despite the obvious lack of progress on certain problems, like housing and salaries, there was generally a feeling that things were getting better after a pep talk from Oppie. In one meeting with young scientists, Harold Agnew ventured to ask why plumbers were getting nearly three times as much pay as staff members with college degrees. Oppie said simply that the plumbers did not know the importance of the Laboratory's work and the scientists did. That, the Director said, was worth the difference in pay.

Who came to Los Alamos? After the war, when asked the question by a Congressional Committee, Robert Oppenheimer said simply, "Everyone." He meant everyone with a reputation in physics. It was true. In two years there was a galaxy of eminent names and Nobel Prize winners parading through the rustic gates of Los Alamos. Joining Oppenheimer as residents or as consultants were luminaries like James Chadwick, Sir Geoffrey Taylor, Enrico Fermi, Niels Bohr, Isidor Rabi, John von Neumann, Ernest Lawrence, and James Franck. The effect was a small university on a mesa with the finest minds to teach and create and the most eager students to learn and work.

Oppenheimer had handpicked most of his key men. Even with his tentative appointment in late 1942, he had some idea of individual men to select to head major research and technical arms of the laboratory. These individuals were old colleagues and friends and had their own reputations in science. In November 1942, Oppenheimer sent General Groves a memorandum on his thinking for key staff that included Edwin McMillan, Marshall Holloway, Joseph Kennedy, Emilio Segré, John Williams, Robert Serber, Eldred Nelson, Stanley Frankel, Edward Teller, and Hans Bethe.[1] Each of these men came to Los Alamos, most to live but some only as consultants. Each man lent credibility to the new laboratory and its work. Their presence also reassured others that there was something promising in the work of that lonely New Mexican outpost.

From the very beginning of the "isolated laboratory," Oppenheimer had had the help of John Manley. An associate professor at the University of Illinois at the beginning of the war, Manley had been drafted from one of Conant's "fission" projects and loaned to Oppenheimer to help plan for the new laboratory and to assist in recruitment. Oppenheimer also wanted both Robert Bacher and Enrico Fermi from the very beginning. Bacher left Cornell to come and brought good men with him. Fermi was still in the midst of reactor research at the University of Chicago, but agreed to come as soon as he could; he would eventually arrive in 1944. Robert Wilson was one of the youngest group leaders at Los Alamos, and brought with him the first large contingent of scientists freed from the cancelled Isotron Project. Wilson was just twenty-eight when he came to Los Alamos. Oppenheimer also personally invited Joseph Kennedy, Robert Serber, Edward Teller, Hans Bethe—all on his "list"—and several others to take major roles in the new laboratory.

Oppenheimer even attracted Niels Bohr, a scientist-god of sorts, who arrived in 1944 with his son, and became known as "Uncle Nick." Bohr was a living legend to most young men at Los Alamos, for it had been Bohr who corrected Rutherford's atomic theory at the age of twenty-eight. He was also the man to whom Meitner and Frisch rushed to tell him that the Germans had discovered fission.

Bohr had barely escaped the Germans in October 1943. Fleeing Copenhagen, he left first for Sweden hidden on board a fishing boat. His young granddaughter followed him hidden in the shopping case of a friendly diplomat. In Sweden, the British flew him to England in the bomb bay of a Mosquito fighter; his oxygen mask did not fit and Bohr arrived unconscious in England. Within a few weeks, Bohr was on his way to America and Los Alamos.

For help with explosives research, Oppenheimer sought help from General Groves. Navy Captain William Parsons eventually arrived on assignment from the U.S. Naval Proving Ground as one of the few men at Los Alamos with previous training and experience in explosives. He assumed responsibility for the Ordnance Division. Within a year, the Laboratory would also add George Kistiakowsky and Norris Bradbury as two others with explosives experience.

The British also sent a team to Los Alamos. The move climaxed two years of negotiations between Churchill and Roosevelt. During 1941, while much of the fission work in America seemed unproductive, the British maintained keen work and optimism in fission bombs. Scientists had adopted the code name "Tube Alloys" for their work. Unfortunately, the war channeled much of Britain's scientific effort into other directions. Reluctantly, Churchill had to admit that only

America had the resources to undertake the research and development of atomic weapons.

Work in several British universities was sufficiently advanced to impress Oppenheimer and his summer team at Berkeley in 1942. Their reports certainly seemed more optimistic than feelings expressed among American scientists. Oppenheimer, with General Groves' approval, then undertook to draft a report to Robert Peierls, the director of the Tube Alloys project in England, in November 1942. The report, however, reflected a hesitancy on Oppenheimer's part to be fully frank. It did not, for example, cover American theoretical speculations on the hydrogen bomb.

Both Roosevelt and Churchill continued discussions throughout 1942 and most of 1943. In the fall, their discussions produced a Combined Policy Committee to be located in Washington. By December 1943, several teams of British scientists arrived in America for placement in various Manhattan District laboratories. In that same month, a team arrived at Los Alamos headed by Otto Frisch and Ernie Titterton.

Sir Geoffrey Taylor also came for long visits at Los Alamos, as did Sir James Chadwick, discoverer of the neutron. Oppenheimer's original thought was that Chadwick would lead the British mission at the Laboratory and have a separate administrative structure to direct. British scientists, however, were needed in different roles in all groups within the Laboratory. Oppenheimer abandoned the idea and eventually British men reported to their American Group and Division leaders. The Mission quickly grew to twenty men. Chadwick returned to Washington and Peierls assumed command. Under him were men with valuable backgrounds in nuclear physics, electronics, and explosives. The group included men like Phillip Moon, Carson Mark, William Penney, and James Tuck.

It was always interesting to new staff to discover the number of foreign scientists at Los Alamos. From Italy had come Fermi, Bruno Rossi, and Emilio Segré; from Hungary had come Teller and von Neumann. Niels Bohr was from Denmark and Stanislav Ulam came from Poland. Both Isidor Rabi and Victor Weisskopf were from Austria, and Hans Bethe and Rolf Landshoff were German refugees. George Kistiakowsky had come to America from Russia as a small child.

Klaus Fuchs was a member of the British team. He was originally a German, but after his escape from the Nazis he became a naturalized British citizen. At Los Alamos, Fuchs was quiet, precise, and a dedicated worker: at times, he would even baby-sit for his married friends. Even before his arrival in Los Alamos, Fuchs had turned over to the Russians American and British plans for uranium separation. Fuchs had acted out of strong sympathy for socialist causes and the belief that atomic secrets belonged as much to the Communist countries as to the Americans or the British. It was not until after the war that everyone learned he had also methodically submitted to Russian agents detailed information and drawings on the Laboratory's implosion bomb.

Several times during his stay in Los Alamos, Fuchs would join his contact, Harry Gold, in Santa Fe. Meeting by the Castillo Bridge, or driving around the outskirts of town in a car, Fuchs would brief Gold on the Laboratory's implosion developments and occasionally turn over a packet of papers written in small, tight script. At the end of their meetings, Fuchs would return to Los Alamos and resume his quiet life. He talked little, remained thoroughly conscientious in his work, and mixed carefully with his colleagues. One wife dubbed him "Poverino"—the poor one.

Unknown to Fuchs—and to government security—the Laboratory had another man at work passing information to the Russians. David Greenglass was a corporal stationed at Los Alamos as part of a "Special Engineering Detachment." He worked at S Site with the design and fabrication of explosive lenses for the implosion bomb. Greenglass had as in-laws Julius and Ethel Rosenberg, and in 1944, the Rosenbergs gave Ruth Greenglass $150 to visit her husband in Santa Fe. During their brief reunion, Ruth suggested that David might pass along information on the Laboratory's work. The information in turn would be passed to Soviet agents. Like Fuchs, Greenglass was sympathetic with Communist philosophy, but unlike Fuchs, he accepted money for his information. By the end of the war, Greenglass had met several times with Julius Rosenberg and Harry Gold to pass along fairly complete sketches of the moulds used to fabricate explosive charges for the implosion weapon.

Both men would escape detection until 1950.

For the most part, it was American scientists who filled the ranks at Los Alamos. Men like Darol Froman, Alvin Graves, Luis Alvarez, Norman Ramsey, Eric Jette, and Joseph Kennedy left established positions at major universities to come to the Laboratory. Kenneth Bainbridge returned from scientific work in England and accepted an offer from Oppenheimer to join the new venture. Norris Bradbury, who would succeed Oppenheimer as director, came from the U.S. Naval Proving Ground, where he had worked with Captain William S. Parsons. Seth Neddermeyer came from the Bureau of Standards. Donald Hornig left the Underwater Explosives Research Laboratory with invitations from both Kistiakowsky and Conant.

There were many more junior men who began their careers at Los Alamos. Harold Agnew, who would also one day head the Laboratory, arrived when he was 22 and just fresh from working with Fermi in Chicago. Oppenheimer seemed equally interested in employing Agnew's wife, who had served as assistant to Richard Doan, director of the Metallurgical Laboratory. Skilled secretaries were highly prized, and she eventually became Robert Bacher's secretary. Recently graduated, Bob Krohn came from the University of Wisconsin with the Van de Graaff machines. Raemer Schreiber left a small Manhattan District project at Purdue University to come. Berlyn Brixner came from an engineering job in Albuquerque. And Richard Feynman, one of the brightest of the young men, came as a new Ph.D. from Princeton to work with Hans Bethe. Even Frank Oppenheimer, Robert's younger brother, joined the Laboratory staff after helping successfully to start the uranium separation plant at Oak Ridge. At Los Alamos, Frank was assigned to Kenneth Bainbridge and preparations for a test of the implosion Fat Man.

It was an extraordinary group. Oppenheimer had managed to attract most of the colleagues whose minds and leadership he valued into a laboratory largely staffed by eager young men. They all had reason to be excited, he declared from time to time. The work was incomparable and the living, he said, in a moment of color, was "charming."

7. ORGANIZATION, 1943

With his appointment, Oppenheimer assumed leadership of a marvelous scientific movement. He meant, as did Groves, to galvanize science behind the war effort, in a way never before imagined. Almost overnight he was thrust from his professorship at Berkeley into directing the most unusual gathering of scientists in the world. It was in the beginning a scientific quest of unparalleled challenge and excitement; in the end there was to be a weapon. The ambiguity of the role of Los Alamos was subtle at first: the form was a scientific laboratory located in the middle of a military post, with heavy secrecy and color-coded ID badges, and the challenge itself was pervasively military in nature.

There was nevertheless an odd beauty and purity to the Laboratory as it took shape in the rustic setting of the isolated mesa. The beautiful New Mexico spring met arriving scientists as the offices of Project Y moved from Santa Fe to the Hill.

Oppenheimer was frantically engaged in efforts to woo America's top scientific and technical personnel into joining what could be called at best a calculated risk. Both Groves and Conant thought it wise to empower Oppenheimer with some statement of purpose. In what became a loose charter for Los Alamos, Groves and Conant wrote to Oppenheimer on February 25, 1943. The letter was to serve as a framework for his responsibilities, and, at his discretion, could be used to solicit staff members.

The letter simply referred to the "development and final manufacture of an instrument of war." Even more ambiguously, Oppenheimer and the new "laboratory" would be concerned with "certain experimental studies in science, engineering, and ordnance." At some future unspecified date, the laboratory would also be concerned with large-scale experiments that involved "difficult ordnance procedures and the handling of highly dangerous material."[1] Oppenheimer became responsible for the conduct of all scientific work and the "maintenance" of secrecy, and was urged to take the advice of his colleagues. Presumably, the letter would assist Oppenheimer in recruiting colleagues.

The importance of close cooperation between the Commanding Officer—Groves—and Oppenheimer was also stressed. This part of the letter clearly set forth the military's dominant role in the project. Oppenheimer had no choice but to downplay the Army's role to some prospective candidates. With only this prospectus and his determination, Oppenheimer set forth to attract the best men he could to Los Alamos.

It was not an easy task. Because of its infancy and nebulous nature, Los Alamos as a scientific establishment did not exist for the academic world. Most scientists with reputations were already engaged in war work at universities. The cadre of physicists and chemists that had worked at Chicago and Berkeley were for the most part already with Oppenheimer or with some other project in the Manhattan District. Although Oppenheimer had few budget constraints—indeed, the urgency gave him carte blanche—his preeminent concern was personnel.

Long after the war, Oppenheimer reflected that "the last months of 1942 and 1943 had hardly hours enough to get Los Alamos established. The real problem had to do with getting to Los Alamos the men who would make a success of the undertaking."[2] His recruitment program therefore had to be massive.

Oppenheimer's concept of a small laboratory had faded rapidly with the growing concerns of the gun and implosion programs. Much later, when the Laboratory's scientific population exceeded 4,000, he noted that he had "underestimated" the size of the Laboratory in the beginning. The primary burden for organizing and managing the

Laboratory nevertheless fell on Oppenheimer. Groves believed that one of Oppenheimer's strongest points was his ability to communicate and work with his colleagues in the scientific world. Between conferences with Groves, overseeing as much as he could the construction of Los Alamos, and beginning work with what small staff he had, Oppenheimer undertook long recruitment trips to exercise his highly lauded skills. "I traveled," he reflected, "all over the country talking with people who had been working on one or another aspect of the atomic energy enterprise, and people in radar work, for example, and underwater sound, telling them about the job, the place that we were going to, and enlisting their enthusiasm."[3]

His enthusiasm was met with skepticism. Despite Oppenheimer's prestige in scientific circles, he was not a Nobel Prize winner, nor, for example, a scientist of the status of Ernest Lawrence. Los Alamos was unknown, located in a place no one had heard of, and although Oppenheimer played it down, it *was* a laboratory on a military post. Moreover, there would be the prospect of being at the project for the duration of the war and severely restricted in terms of travel and family freedom. And there was the very real spectre that the work might fail and perhaps, with that, a worker's reputation. It was not a challenge for the timid.

One of Oppenheimer's first tasks was to organize the Laboratory administratively. Generally, he wanted to follow the university model of scientific divisions organized around a central administrative hub. He made the division of technical responsibilities among four programs: Experimental Physics, Theoretical Physics, Chemistry and Metallurgy, and Ordnance. Each was to act as an administrative division which in turn was to consist of operating units or groups. Group leaders were responsible to division leaders, and division leaders were responsible to Oppenheimer.

To handle administrative matters for the Laboratory, Groves pressed Oppenheimer to hire Edward U. Condon from the Westinghouse Research Laboratories. As the associate director, Condon was to follow the general recommendations of the April Review Committee in providing Oppenheimer with relief from purely administrative matters. The position would also serve as liaison with the military commander of the Los Alamos Post. Condon clashed with both military and scientific personnel almost immediately, and stayed only six weeks. Groves had to assume responsibility for the disaster. Condon, Groves said, "did little to smooth the frictions between the scientists and the military officers who handled the administrative housekeeping details."[4] Although an intelligent man, Condon was unable to bring together the Post military personnel, the Army Engineering Corps, and the civilian residents of Los Alamos. When he resigned, Groves feared that his release might spark trouble at some future date. General Groves encouraged Oppenheimer to request a written letter of resignation from Condon. Groves suspected that Condon left because he feared the project would fail and damage his reputation.[5] It took Oppenheimer six months to find a suitable replacement. In the interim Oppenheimer appointed special assistants to take charge of administrative duties.

Groves would have preferred to have all of Los Alamos in military uniforms. Security needs alone seemed to require military control. Conant, who had had experience in military laboratories during World War I, expressed no particular objection to the idea. Preliminary thinking called for Oppenheimer to be made a Lieutenant Colonel and all division leaders to be Majors. Immediate objections were raised by other scientists. Groves was told that placing scientists and technicians as officers and enlisted men would not be conducive to open work and decision making. Moreover, recruitment would suffer, perhaps fatally, if participation in the project meant being in uniform. It was strongly suggested that a military organization would be rigid and inimical to a free exchange of ideas. How, for example, would an Army officer admit to being wrong or change a decision?

Oppenheimer did not immediately object, however, and took preliminary steps toward becoming an Army officer in San Francisco. Key men, such as Isidor Rabi and Robert Bacher, strenuously objected to a military laboratory. Their objections helped turn the tide toward a civilian effort. Oppenheimer desperately needed their participation in the project. Their objections were a real obstacle to further recruitment. Conant was consulted and finally he and Groves were persuaded to maintain the Laboratory, for the moment at least, as a civilian enterprise. Such a promise was incorporated into the February 25 letter to Oppenheimer to reassure individuals during recruitment. Both Groves and Conant reserved the possibility of future military conversion. The period until January 1944 would be civilian, however.

Los Alamos would nevertheless be placed on a military post with its commander reporting directly to General Groves. Although Oppenheimer and the others had gained only a partial ten-month victory, it was enough to enable recruitment to proceed unhampered. In the end, no military takeover occurred, and the Laboratory remained civilian.

With a general administrative structure in mind, Oppenheimer arrived in Santa Fe on March 15 to spend the next two weeks organizing the few men who had arrived with him. The largest contingent came from Princeton University, where the cancellation of Robert Wilson's "Isotron Project" freed the largest block of scientists for transfer to Los Alamos. The failure of Wilson's electrical uranium separation project released some thirty people. The remaining core of early staff came from Oppenheimer's scattered "fast-neutron" projects at Berkeley, University of Minnesota, Stanford, and Purdue. A few men were freed from work at the Metallurgical Laboratory in Chicago and encouraged to go to Los Alamos; others were recruited from M.I.T., Columbia, Iowa State University, and the National Bureau of Standards.

With the arrival of more men, Oppenheimer relinquished his preliminary directorship of the Theoretical Division and formally placed Hans Bethe in charge. Born in Germany, Bethe was now an American citizen and came from the Radiation Laboratory at M.I.T. Robert Bacher had been involved in the development of radar at Cornell and came to assume leadership of the Experimental Physics Division. U.S. Navy Captain William S. Parsons took directorship of the Ordnance Division. His experience in the development of radar and the proximity fuze was valuable, but his knowledge gained from work with explosives at the U.S. Naval Proving Ground was critically needed. Joseph W. Kennedy had been a student of Glenn Seaborg and assumed the leadership of the Chemistry and Metallurgy Division. On administrative matters, Oppenheimer replaced Condon with two men. David Hawkins came from the University of California to act as administrative assistant to Oppenheimer and as liaison between the Laboratory and the military post. A. L. Hughes, formerly chairman of the Department of Physics at Washington University, was asked to become personnel director.

Oppenheimer diffused the Laboratory government by creating a Governing Board to act as counsel to him. This body would have as members the Laboratory director, all division leaders, and administrative and technical liaison officers. In addition to acting as counsel, the board had two specific functions: to plan and conduct the technical effort, and to act as a directorate in which Oppenheimer and his principal staff could collaborate in decision making.

The first members of the board were Oppenheimer, Robert Bacher representing the Physics Division, Hans Bethe from Theoretical, Joseph Kennedy from Chemistry and Metallurgy, William Parsons from Ordnance, and Llewelyn Hughes and Dana Mitchell representing the administrative offices. Later, Edwin McMillan, George Kistiakowsky, and Kenneth Bainbridge were added. The first few meetings were hardly concerned with research and development. Instead, the board dealt with housing, construction problems, security, morale, and promoting policies.

Shortly after the creation of the Governing Board, Oppenheimer also created the Laboratory Coordinating Council. The membership of this body was primarily group leaders or personnel above that level. The council was not a decision-making body such as the Governing Board, but could have policy matters referred to it. Oppenheimer announced this new body at the June 17 Governing Board meeting. Robert Bacher was asked to prepare an agenda, and either he or Oppenheimer was to preside over the meeting. The original membership consisted of twenty members.

In June a third group was added to the Laboratory organization. Suggested by Hans Bethe, the Colloquium was created as a forum for exchanging ideas among scientists. Membership was limited to "staff members," individuals with scientific degrees or equivalent training. These individuals, in contrast to other Laboratory staff, were regarded as "contributors to or beneficiaries from an exchange of information." It was also a vehicle for developing a sense of common effort and responsibility in which younger staff could become equals with older, more senior men.

General Groves was shocked at the prospect of such openness. It was antithetical to compartmentalization, and worse, Groves thought, an open invitation to security leaks. Regular attendance would give any participant a complete view of the project. This was its purpose. Any individual would be able to participate and contribute. Groves objected, but Oppenheimer stood firm. It was critical, he said, that everyone have the chance to contribute.

To placate Groves, however, Oppenheimer agreed to limit membership and place all participants on a voucher system. In practice, the colloquia were often broad and academic in content and sparse on detail. It was a major departure from military custom, but well worth the risk in developing an esprit de corps at Los Alamos. The Governing Board set the first agenda for the Colloquium on May 30: a discussion of the three or four most promising methods of uranium and plutonium production, questions of measurements, and various ordnance characteristics of explosives.[6]

Groves was still unhappy. In late June he raised the question of the Colloquium with his Military Policy Committee. While they could not dispute the potential value of the colloquia in nurturing every resource within the Laboratory, they did feel compelled to urge tight control. Bush was then asked to persuade Roosevelt to write to Oppenheimer with cautions on security. By chance, Roosevelt called in

Bush for a review of the fission project and Bush was able to suggest that a letter be sent to all leaders with the Manhattan apparatus. Roosevelt concurred. A letter was sent to Oppenheimer encouraging the work of Los Alamos and imploring the need for security and for living under unusual restrictions. Roosevelt closed with characteristic optimism. American scientists, he said, would be equal to the challenge.[7]

Oppenheimer had more than security to contend with. Although Los Alamos had been given the task of developing a weapon, it was at the moment heavily dependent on other Manhattan projects. Not only did the Laboratory depend on the huge plants at Hanford and Oak Ridge for raw nuclear materials, but also on the smaller laboratories, such as the Metallurgical Laboratory at Chicago. There was a great need for information from these other sites. The shroud of secrecy between Manhattan laboratories was very frustrating.

The question of liaison and information exchange came into partial resolution in June. Groves thought the demands of security so great that only the most critically needed information should be exchanged. He drafted a set of procedures for liaison: exchanges, for example, between Los Alamos and Chicago were limited to correspondence between specified representatives and restricted to chemical, metallurgical, and certain properties of nuclear materials. No information could be exchanged on the design of weapons, the operation of piles, or comparison of schedules.

Groves relented further when Oppenheimer convinced him to approve a personal visit to Oak Ridge. Oppenheimer felt it urgent that some information be obtained from Tennessee on U 235 production. Moreover, scientists at Los Alamos needed to know what form the uranium would arrive in and what kind of processing it had undergone. On July 2, Oppenheimer wrote Groves to insist that all division heads in the Laboratory be kept "adequately" informed on production schedules.

Ironically, the issue of information between projects became less important as the Manhattan projects matured. Los Alamos, particularly, benefited from the growth of the Manhattan effort. As personnel needs grew within the Laboratory, scientists and technicians were transferred to the Laboratory from other sites. They brought with them details on work at previous projects. Los Alamos was able to piece together more information than Groves would have liked.

But normal information and business channels were complex. Los Alamos concealed its existence and the identities of its staff when making inquiries of agencies outside the Manhattan District. Blind addresses were used; long-distance telephone calls were made through a telephone number in Denver; and government identification cards were specially made with false names and serial numbers. The Laboratory was forced to rely heavily on its offices at the University of California for ordering and procurement. This dependency almost always caused delays and errors.

Many complaints inevitably found their way to Oppenheimer. Condon's arrival and abrupt departure did little for civilian-military relations. The Army, and particularly the Army security men, were saddled with the task of administering a civilian laboratory haphazardly placed within a military setting. Most of the Army staff were as much in the dark about the work at Los Alamos as anyone else. Most of the responsibilities were certain to engender problems in human relations: security, censorship, transportation, commissary privileges, and the like. Groves was perpetually distressed over what he regarded as the prima donna behavior of the scientists and their families. On the other hand, Groves appeared to the scientists as the very epitome of the insensitive, anti-intellectual military mind. Privately, Groves referred to Los Alamos civilians, young and old, as children; General Groves was called many things by Hill residents, including "Goo-goo eyes."

The relationship somehow remained workable, even in matters of security. Oppenheimer himself chose to be thoroughly cautious and tried to keep a tight rein on security. In May Oppenheimer met with his Governing Board to discuss the creation of a story to explain the work of the Laboratory to outsiders in the immediate area. He was also sensitive to the inquiring and loquacious nature of his colleagues and sought to prevent stricter security measures from Groves by stressing among his staff the need for obeying what rules were at hand.

The ever-vigilant Groves continued to press security and to bring staff infractions to Oppenheimer's attention. From the very beginning at Los Alamos, Groves had ordered all ingoing and outgoing Laboratory communications monitored and censored, where possible. As early as February 1943, the Los Alamos Post Commander had been ordered to exercise "complete censorship" of communications in order to maintain the secrecy of the project.[8] This task became more difficult as Los Alamos grew and increased its contact with other Manhattan District projects. Groves continually found it necessary to remind Oppenheimer to "tighten up" security, and frequently wrote terse letters citing specific security breaches. Groves was particularly disturbed by the careless use of the telephone by scientists. Parsons' men working with Kingman—the code name for Wendover Field in Utah—came under particular fire in a burst of several letters from Groves to Oppenheimer. Groves even named specific men who had been secretly monitored over telephone lines tapped by Post Intelligence.[9] Oppenheimer passed the word down and again urged his senior leaders to exercise stronger control over their men.

Oppenheimer was remarkably successful in urging his colleagues to obey the rules, in spite of difficult living conditions. There was a constant succession of everyday problems in Los Alamos: housing shortages, little water, poor roads, irregular supply of food items in the commissary. Against these problems, Oppenheimer continually had to persuade staff to work full force, keep their own families happy, and somehow maintain cheerfulness and dedication.

One particular problem plagued Oppenheimer throughout the war: salaries. Initially, two standards for salaries were employed. The first was the OSRD* scale, which was based on the scientific degrees held and the number of years since the individual had earned them at his university. The second standard was based on a "no-loss, no-gain" principle, in which individuals from academic settings who were paid on a nine-month basis were now paid at 12/10 their previous rate. There were inherent difficulties, because men from industry were usually better paid than their academic colleagues. Another problem was that technicians—men usually with no formal degree but with considerable technical skill—had to be paid at rates consistent with the current labor market. In numerous cases, technicians who ranked lower in responsibility at Los Alamos were paid more than younger scientists. Because of its youth, the Laboratory had no formal system for granting service or merit increases.

Oppenheimer finally asked Hughes, as personnel director, to prepare some sort of policy for the Laboratory. Hughes in turn devised a plan based on OSRD scales and submitted it to Groves and the University of California as prime contractor. Action was bureaucratically delayed in Washington until February 1944. During this time there were no promotions. Although other plans were drawn up and used to one degree or another, no formal system emerged until the end of the war.

Housing, security, salaries: these were all problems that annoyingly hid what Oppenheimer believed to be the truer nature of the Laboratory. Oppenheimer envisioned Los Alamos with a measure of hope and metaphysics: true scientific spirit, daring challenges locked in nature, and the interplay of minds. Administration only made that interplay possible. Oppenheimer was not ignorant of human nature,

---

Office of Scientific Research and Development.

and took on administration as best he could in order to provide the sort of atmosphere he sought for his staff and their mighty tasks. As a result, administration always ran second to science.

Groves did not share Oppenheimer's belief in the intrinsic and sometimes ethereal beauty of science. He was interested in putting science to only one pragmatic test: development of a bomb. Oppenheimer needed to remember, as one military man suggested to him, that after all it was just a product with a bang.

8. THE FIRST YEAR: 1943-1944

Los Alamos was a gamble. Groves pushed and bullied his Manhattan apparatus with a ferocity that often angered civilians and military alike. Oppenheimer was more aware of the gamble and was more gentle, more conciliatory and understanding; he urged where Groves ordered. Oppenheimer and his colleagues at Los Alamos knew that basic research promoted the possibility of an atomic weapon. Fermi's successful chain reaction at Chicago confirmed that the basic nuclear process in fission was the same process that would create an explosion under the right circumstances. But what were they?

There were fundamental unknowns that affected the entire project. No one knew for sure, for example, whether fissioning plutonium 239 emitted enough neutrons for an explosive chain reaction. A knowledge blank in one area meant that work in another area had to wait. During the first few months, many groups within the Laboratory were waiting on one another for baseline information that would allow them to proceed with their own experiments. The first year was to be one of heavy research in which the "pure" scientists were preeminent; during the second year, their role would decline in favor of the military scientists.

There were two basic types of weapons envisioned: a uranium and plutonium gun bomb, and an implosion bomb using either element. Most work presupposed the use of either uranium or plutonium, but a small task force within the Laboratory set about exploring another form of uranium called uranium hydride. Eventually the hydride possibility was dropped because its explosive yield was suspected of being very small. The gun was the best bet. It incorporated a simple design in which one piece of active material was shot within a closed barrel into another piece at the opposite end. In the gun model, uranium differed from plutonium only in the speed with which it needed to be shot into the second half. Uranium required velocities around 2,000 feet per second, while plutonium required almost 3,000. The higher speed was necessary because of the high neutron "background" in plutonium. If it were fired at a slower speed it could not connect with its other half in time to prevent predetonation.

The implosion bomb was a far more difficult design, both in theory and in technology. The concept of implosion had been recognized as early as 1942 and covered in Tolman's report to Roosevelt. By compressing active material symmetrically and with great force, the material could achieve supercriticality. There were two advantages to implosion: such a bomb could use plutonium—which promised to be available perhaps before U 235—and uranium, and the fission process would work at a much greater speed. This last advantage meant that the assembly of all components occurred quickly and lessened the chance of a misfire.

Oppenheimer was under great pressure to weigh all the factors in both approaches. Both he and Groves were committed to pursuing the course that would most rapidly lead to success. There were many obstacles to developing an implosion weapon. Implosion was still basically theory. Los Alamos as a staff had very little experience with conventional explosives and none with unorthodox ones. Not enough plutonium was immediately available to conduct critical experiments and to obtain fundamental knowledge on the metal's nuclear properties. While plutonium production seemed very promising, the plant at Hanford was just under construction, and major delays in delivery might make all plutonium weapons problematical.

Just how much U 235 or Pu 239 it would take to create an

explosion was not known for certain. On February 4, 1944, Groves wrote Roosevelt that this question was receiving urgent attention at Los Alamos. The best guess, however, was that from 8 to 80 kilograms of U 235 would be necessary for the gun weapon. The question of plutonium, however, was not raised in his letter.

Oppenheimer polled his men. He was undecided as to how much support to allot to implosion. Clearly he could not jeopardize the gun program, which promised the most easy and quick success. Still, his Theoretical Division continued to argue in favor of the concept, and Seth Neddermeyer continued to pose intriguing questions. At several Governing Board meetings in September and October, Oppenheimer threw open the discussion. The consensus seemed to suggest that only limited work could be encouraged. Oppenheimer took the discussions to heart. Implosion research would continue, but the gun bomb was the first priority. The majority of the Laboratory's resources were to be thrown behind it.

By necessity, work on the gun program was diffused throughout several divisions of the Laboratory. Bethe's Theoretical Division began research into two key elements of bomb development: critical mass and the so-called nuclear efficiencies. The division reflected Bethe's personal style, and was loosely organized until early in 1944. At that time there was a small reorganization within the division: Edward Teller was taken off gun research and left to pursue theoretical work on the implosion and thermonuclear bombs. Robert Serber was assigned diffusion experiments concerned with determining the distribution of neutrons in a given nuclear mass. Victor Weisskopf took on nuclear efficiency, and Richard Feynman and David Flanders divided diffusion and mathematical work.

The most immediate task was to calculate what scientists were calling nuclear efficiency. This concept involved understanding the nature of a chain reaction and how the neutrons operated within it. The entire process was incredibly difficult to grasp, because during the cycle from detonation to supercriticality many properties of the components—such as the core and tamper—were being intrinsically transformed by the terrific forces inherent in the cycle itself. It was as if an unknown machine could be understood only if each part was separately identified and its relationship to all other parts clearly known. Others in the Laboratory, such as Parsons' men in ordnance, were waiting for this information in order to undertake the preliminary design of bomb components.

Critical mass was still another unknown area. At the time the division was formed, there was little nuclear material for use in experiments. The division had to continue with its best guesswork. A critical mass was the smallest amount of fissionable material—either U 235 or Pu 239—that could support a chain reaction. For purposes of the Laboratory, the men under Bethe had to determine what amount of material, along with explosives, tampers, and initiators, would cause an atomic explosion.

Despite the pressing needs of the division, Teller insisted on studying the superbomb. Teller had remained fascinated by its possibilities since a lunchtime discussion with Fermi in 1942 had brought up its potential. The concept presented to the Laboratory in April involved the detonation of liquid deuterium (heavy water) through the ignition of a fission bomb. Teller and his staff of two others realized that the superbomb was more of a possibility when the artificially produced element tritium was added. This suggestion by Konopinski, a member of Teller's crew, meant that the ignition temperature of deuterium could be lowered. With a lower ignition point, the probability of a man-initiated explosion increased.

Teller pushed Oppenheimer and the Governing Board in September 1943. He reported on the enhancing qualities of tritium and suggested that the recently reported rumors of German interest in deuterium was *de facto* evidence of German work in superbombs. The "super" was clearly outside the Laboratory's main drive toward the gun and implosion bombs. The Board stalled and took up the question again in February 1944. Groves had become more interested, however. He sent Tolman as his representative to meetings at the Laboratory in February. Tolman offered mild support for Teller. Such a weapon needed study, he argued, but it appeared likely that it could not be developed in time for use in the war. Moreover, it could not deter fission work. Teller had to be content, for the moment at least, with his independence and a small staff to continue work on thermonuclear weapons.

A building away from Bethe was Robert Bacher's Experimental Physics Division. This was one of the first units to be organized at Los Alamos; Oppenheimer himself had taken great care to recruit Bacher as its head. Bacher, Bethe, and a few others were asked to be Oppenheimer's key staff. Bacher accepted Oppenheimer's offer and organized his division into seven groups: Cyclotron, headed by Robert Wilson; Electrostatic Generator, headed by John Williams; Neutron Source Research, led by John Manley; Electronics, led by Darol Froman; Radioactivity, led by Emilio Segré; Instrumentation, or the "Detector Group," headed by Bruno Rossi; and the Water Boiler, or Reactor Group, led by Robert Christy.

The first contingent of the division arrived in Los Alamos in March 1943. They were met with confusion, and had nowhere to set up their equipment. Nothing was complete. Equipment was arriving daily from laboratories at Harvard, Princeton, Wisconsin, and Illinois. Smaller equipment was held up until the cyclotron, van de Graaff, and Cockcroft-Walton generators were installed. Only slowly could these huge pieces of equipment be set up, checked, rechecked, and finally put into operation. The Harvard cyclotron found a permanent home in April in Building X; the two van de Graaff generators from the University of Wisconsin were installed on the ground level of Building W, and the Cockcroft in Building Z.

With minimal equipment in place, the division set out to measure the time it took for neutrons in a fissionable mass to generate additional neutrons in a chain reaction. Bacher's men had only one extremely thin piece of U 235 metal foil with which to conduct experiments. That foil represented the entire stockpile of precious uranium available at the time. Even a year later, with the supply of U 235 up to a few pieces of foil, the Laboratory had managed to acquire only a few specks of Pu 239.

The question of whether sufficient neutrons are generated during the fissioning of uranium and plutonium was most important. An explosive chain reaction could be achieved *only* if sufficient neutrons were generated. In July, the van de Graaff staff under John Williams was able to test a microscopic speck of plutonium brought to Los Alamos by Glenn Seaborg in his suitcase. Everyone was vastly relieved to learn that plutonium re-emitted neutrons in numbers slightly greater than U 235. This discovery meant that plutonium could be used in a weapon. Groves was pleased for a second reason: he was delighted to learn that the discovery justified the massive plutonium pile under construction at Hanford. Always conscious of his role as fiscal administrator, Groves sought as best he could to have his expenditures justified at every turn.

In November Wilson's men were able to confirm that most of the neutrons emitted from the fissioning of U 235 were "fast" neutrons, and thereby confirmed that success of a gun weapon was a certainty. The speed of emission was outstanding: neutrons were generated in less than a billionth of a second.

The division also took up the concept of the tamper, or reflecting mechanism, that would surround the uranium and plutonium and bounce back neutrons into the core. Manley and his team ran numerous experiments on metal discs and large spheres made of a dozen metals: lead, gold, platinum, tungsten, uranium, and others.

Vistitors to Manley's office would occasionally stumble on a metal disc used as a door stop; they were always surprised to learn that it was made of solid gold.

On his arrival, Bacher began plans to locate and construct his atomic reactor, the "Water Boiler." There was only Fermi's Chicago experience to go on, but great care was given to locating the reactor so that it would be accessible but relatively harmless to Los Alamos if something went wrong. The original plans called for a boiler that would operate at about 10,000 watts and produce some 3,000 curies of radiation. The "curie" was named after Marie and Pierre Curie, the discoverers of radium, and was the basic unit used to describe the intensity of radioactivity. In the case of an explosion, or a runaway chain reaction, the radioactive damage could be severe. The division finally chose Los Alamos canyon below the mesa and called it Omega Site. The road down from the Hill used part of the old boys' school road that clung to the right side of the canyon wall.

Fermi came from the University of Chicago in September 1943 to consult with Bacher and Robert Christy's Boiler Group. He had several problems waiting for him. There was a strong suspicion that the reactor would produce a gas that would hamper or prevent operation. No one knew what to do with the radioactive residue that would form after the reactor had been in operation for a few months. Fermi finally suggested heavier shielding around the reactor, and strongly urged that the reactor's operation be cut back considerably. There was disappointment. Reduction of power would limit the boiler's usefulness in a wide range of other experiments. There was little that could be done, however, and in May 1944 the boiler sustained its first successful but limited chain reaction.

Joseph Kennedy had the work of his Chemistry and Metallurgy Division doubled following the recommendations of the April Reviewing Committee. In addition to the basic chemistry and metallurgy work that had been anticipated, Kennedy now had to undertake the purification of both uranium and plutonium. His division had hardly enough space. Within months his buildings began to spread across the mesa. Dominating the area was the new massive D Building for working with plutonium.

For almost a year, the division was loosely organized around its three primary functions: purification, radiochemistry, and metallurgy. By April 1944, Kennedy was forced to respond to Laboratory-wide pressure to provide additional services. His expansion and reorganization included new groups to deal with health and safety, high vacuum research, and the analysis and recovery of uranium and plutonium. Groves provided some relief by shifting to Los Alamos the production and purification teams of the Metallurgical Laboratory and of the chemistry laboratories of Berkeley and Iowa State College. In May 1943, Oppenheimer asked Charles Thomas of the Monsanto Chemical Company to come and coordinate the activities of the division and all other related Manhattan efforts. Thomas came in July and assisted in the design of the Laboratory's metallurgical facilities. Most of the buildings were indistinguishable from other Los Alamos buildings. They all kept the same drab exterior with green roofs. The chemistry and metallurgy buildings did employ special construction that included air conditioning and special ventilation systems to eliminate dust and small particles from the air.

All of the Laboratory grew wildly but none faster than Kennedy's division. In June 1943, the division had only 20 staff members on board; by the end of the war, the staff had grown to over 400 scientists and technicians. Men were drawn from all available sources and still the demand rose; finally, in 1944, Oppenheimer and Kennedy were forced to ask the Army to hurriedly reassign men whose talents were needed at Los Alamos.

The most extensive program in the division was metal purification. The purity of uranium needed for a bomb was less than that of plutonium. In fact, scientists estimated that uranium might be less than one third as pure as plutonium.[1] Kennedy therefore directed his staff to concentrate on developing techniques for purifying plutonium. Later, in 1945, this was proved to be a very wise decision.

The uranium purification process was a combination of what were called wet and dry stages. The wet stages involved obtaining precipitants through chemical procedures and then, in the dry stages, igniting the substance, oxidizing it, and heating it in the presence of several other chemical compounds. The plutonium purification process also involved wet and dry stages. Recovery steps were incorporated into work with both metals because of their extreme scarcity.

When the plutonium first began to arrive in small quantities from Clinton in 1944, and later from Hanford, it was a viscous mixture of decontaminated and partially purified nitrates. The plutonium had to be chemically dissolved out of its stainless steel shipping container, diluted, and a sample extracted for radioassay. This process alone took three or four days and prepared the metal for the wet purification steps.

Working with plutonium presented many hazards. Kennedy shared responsibility with Dr. Hempelmann, the Laboratory's medical director, for plutonium and uranium health concerns. Both men became involved with methods of preventing plutonium poisoning, as well as finding means for detecting it in the human body. The extremely fine physical quality of the metal made it very easy to ingest into the body by breathing. Often enough an individual would be exposed to plutonium, but it would be difficult, if not impossible, to tell to what extent.

There was even some question of what amount constituted an overdose. Hempelmann believed that the most satisfactory method of testing for exposure was by examining the excreta. This process seemed about 90 percent effective. Nose dabs to check inhalation were another possibility, but both Kennedy and Bacher felt that examining the lungs would prove questionable as a test because of uncertainty as to how plutonium remains in the lungs.[2] Their concerns and approaches were novel, of course, but so was the state of the art. No one had ever worked before with a substance that didn't exist in nature.

In June 1943, the Ordnance, or E Division, occupied only three small rooms in Building U, directly behind Oppenheimer's office. Both Bush and Conant wanted Captain William Parsons to head this division. With the approval of Groves and the Governing Board, he was transferred to Los Alamos from the Navy in June. Parsons picked his men carefully. Edwin McMillan took over the explosives testing ground; Kenneth Bainbridge, just back from research in England, assumed leadership of the instrumentation group; Robert Brode led fuze development; Charles Critchfield assumed the responsibility for the ominous-sounding Projectile and Target Group; and Seth Neddermeyer led a small group of men on implosion studies.

Like most divisions, Ordnance expanded quickly. Results from experiments by the Theoretical and Experimental Physics divisions came directly to Parsons and his teams. The purified uranium and plutonium metal would eventually be given to the division for assembly into weapons. Parsons quickly added men and additional organizational units. By the end of 1943 he had shuffled his division to include three new groups called Engineering, High Explosives, and Detonators.

Oppenheimer noted the rapid expansion, and met with Parsons to appoint two new deputy division leaders. Edwin McMillan took over direction of the uranium and plutonium gun programs, and Kiev-born George Kistiakowsky was asked to come from Harvard to lead the implosion program.

For most of the first year, the division's energies were directed toward developing the gun bomb. Several gun systems were considered. Teller, for example, advocated the use of two guns, to be fired from different locations, with subcritical masses in each. The two guns would be mathematically programmed to have both subcritical parts meet in midair above the target; as they met, they would become supercritical and explode. Someone else suggested the use of jet propulsion. The double-gun system was soon dropped, in favor of concentration on developing a single gun weapon.

As Parsons assembled his team in 1943, it was obvious that there were huge gaps in what they needed to know. Only the roughest estimates could be made on the amount of active material to be fired, the speed required to bring the parts together, and in the case of plutonium, how much acceleration the metal could withstand and still become supercritical upon contact. Although the gun bomb employed conventional ballistic characteristics, it was also very much unlike traditional weapons. There was no concern, for example, over stability in flight; there was no energy to absorb in recoil; and the gun barrel itself could be lightweight, because it would be used only once. The plutonium gun—if it would work—required even higher velocities, and the only weapon that reached 3,000 feet per second had proven unreliable and was abandoned as a model.

Parsons had to design a new gun for use with plutonium. The gun was to weigh only one ton instead of the usual five tons for a gun of similar muzzle velocity. It was to be made from highly alloyed steel. The gun would also be as short as possible and employ three independent primers; the maximum pressure at its breech was to be nearly 75,000 pounds per square inch. There was some difficulty in finding a firm to make the weapon, but two gun barrels finally arrived in March 1944.

While waiting for the new guns to arrive, Parsons put his staff to designing and building testing installations for explosives. Their plans called for a conventional testing range located about five miles from the Tech Area and called Anchor Ranch. The buildings included gun emplacements, sand butts, and bombproof control rooms and magazines. When the size of the testing expanded and requirements became better known, more reinforcement was added to the walls and roofs and some buildings were placed in ravines for better protection.

Other members of the Ordnance Division were concerned with arming and fuzing the bomb. Led by Robert Brode, the Fuze Development Group had to find a dependable way to fuze the new weapons. Triggering mechanisms that failed even one percent of the time were unacceptable. The available supply of uranium and plutonium was so small that triggering had to be nearly perfect. Brode's problems were complicated because plans in 1943 called for the bomb to be detonated several hundred feet off the ground. There were no fuzes commercially available for this purpose.

There were two approaches to firing studied by Brode's team: currently available barometric switches would be adapted, or more complex devices such as the radio proximity fuze, radio altimeter, or tail-warning device could be perfected and used. Clocks were even a third, distant possibility, but were not the best choice because human error might cause failure to make the right setting just before bombing. Barometric switches were the simplest device, but there was no assurance that such a fuze could be depended on in a falling bomb. The group began a wide series of experiments with model bombs outfitted with radio transmitters and dropped from airplanes to check barometric readings and actual elevation. When the actual dimensions of the bomb casing were known in 1944, another series of mock-up bombs was dropped at the Muroc Army Air Force Base in Utah. All the evidence suggested that the barometric fuze should at best be a backup to a more reliable method.

Brode had already made early estimates of the required height for explosion. Generally, the altitude would be below 500 feet and more probably below 150 feet. The proximity fuze was already well developed and tested for use by the military; there was a great need for accurate detonation of bombs dropped by airplanes. The Laboratory, along with the University of Michigan, began a development program to test the proximity fuze. Word in February 1944 suggested that the bomb could now be detonated at heights as high as 3,000 feet. The proximity fuze was ineffective at these heights and Brode was once again forced to look for new electronic methods.

Independent of Los Alamos, the Radio Corporation of America had developed a tail-warning device that appeared to scientists to be adaptable to Laboratory needs. A substantial number of RCA's pilot models were made available to Brode for experimentation. The units were quickly nicknamed Archie by Laboratory staff. Tests with actual airdrops of Archie seemed to strengthen the belief that this new device was the answer.

While Brode and his men struggled with the fuzing and detonating systems, Norman Ramsey was asked to take on a new group called Delivery. Its purpose was to process the bomb from its final Los Alamos shape into a weapon that could be used in the field and air-dropped on a target. Ramsey began in the summer of 1943 to investigate military planes that might be adapted to drop a fairly large payload. The plutonium gun was estimated to be some 17 feet long; only a B-29 could handle such a bomb size and then only so by joining both bomb bays together. The British Landcaster bomber was rejected for political reasons, despite the possibility that a B-29 might not be immediately available for testing purposes.

Ramsey couldn't wait for a new B-29 to begin testing. Conventional airplanes were used to drop scale models called the "long thin man." These models were 14-inch pipes welded to the middle of a standard 500-pound bomb. In late September 1943, Ramsey took data from the Theoretical Division and calculated an implosion weapon to be a large round bomb some 60 inches in diameter. Fins would be added for stability. The whole device was nicknamed Fat Man. Almost a year later, Ramsey developed full-size mockups of the Fat Man and began a series of test drops in Utah.

Parsons found himself in the odd position of designing field weapons whose parts were not all known or understood. In the beginning, particularly, their work seemed as much luck as anything else.

The Theoretical Division made an important discovery in the summer of 1943: a large part of the kinetic energy of an "imploding" explosion would be transferred to potential energy or compression. Almost incompressible solid matter would be compressed even further. The discovery added further strength to the implosion movement at Los Alamos.

For almost a year, the only work at the Laboratory on implosion was within the Theoretical Division. Both Teller and Rudolph Peierls of the British Mission spent considerable time trying to develop some understanding of how the explosive charges would have to work. The mathematical calculations involved were exceedingly long and complex, and were largely done by hand on notebook tablets. In April 1944 the division received a godsend in the shape of an early set of IBM calculators. Rudimentary as the machines were, they expedited the mathematical work enormously. Very quickly there was some limited grasp of the explosive dynamics involved: all explosive charges would have to detonate not only at the same moment, but also with complete uniformity around the plutonium or uranium core. James Tuck, a young scientist in the division, suggested that the explosive force could be facilitated by utilizing charges shaped much like blunted pyramids. This configuration would assist in forcing the shock waves to converge uniformly on the active material.

Tuck's idea was novel, but it meant more research into what happened when heavy and light materials were forced against each other during an explosion. It seemed clear that such an explosion must produce stability or it would not work. Heavy material must push against light for stability. In the case of implosion, the charges must be correctly adjusted for the weight and quality of the tamper and core. This symbiotic relationship demanded that much of the Laboratory's implosion research be directed at understanding the appropriate shapes for the explosive charges.

Despite the theoretical and technical difficulties, implosion continued to gain support in the Laboratory as the year wore on. Seth Neddermeyer led what at first was a small implosion team in Parson's Ordnance Division and started working in May with Teller on various theoretical aspects of the process. Neddermeyer was the earliest champion of the implosion bomb and one of its heartiest defenders during the months when Oppenheimer and most of his leaders saw only the gun weapon completed in time for use in the war.

Implosion as a serious alternative had gained sudden momentum in the summer when a disquieting word was passed from Compton in Washington to Oppenheimer in June 1943. Compton sent a report compiled by Richard Tolman which reported the most recent work by Frédéric Joliot in France, where the scientist had noted a spontaneous emission of neutrons from fissioning polonium, an element lighter than uranium or plutonium. This discovery had momentous implications for the work at Los Alamos.

Joliot reasoned that his discovery of spontaneous emission might also be expected from plutonium. Fermi, who had read the report in Chicago, also believed that this would be the case. Oppenheimer urgently called his Governing Board together on June 17 and anxiously read the report.

There was great reason to be excited. If Joliot's findings were accurate, it was almost certain that plutonium would act in the same way and therefore could not be used in a weapon of the gun-assembly type. Spontaneous fission was a great possibility because of the neutron background. Such fission would cause predetonation and prevent a nuclear explosion. The gun assembly method for plutonium would fire too slowly to overcome predetonation and would simply fizzle. Only the implosion bomb could utilize plutonium.

There was incredulity at Los Alamos. Bethe suggested that such a reaction might be due to impurities, and after long discussion, Oppenheimer asked Emilio Segré to assemble a staff in Pajarito Canyon to explore the possibility.[3] Unfortunately, there was no immediate way to test this condition. No one could be sure that spontaneous fission would occur without a sample of plutonium. Los Alamos could not expect any of this substance for almost a year. And whatever plutonium would arrive from the Clinton reactor could not be expected to be entirely free from impurities. Without empirical confirmation of this gloomy possibility, Oppenheimer had no choice but to pursue the gun weapon as planned. He could, however, allow Neddermeyer more men and resources to develop the implosion alternative.

Neddermeyer was thoroughly convinced of the workability of the new weapon. He had come to Los Alamos from the Bureau of Standards as a former student of Oppenheimer's and on the director's personal invitation. Once at the Laboratory, he had been given a small staff but wide latitude to explore implosion. Unfortunately, his shy manner and poor management style soon fell behind his increased responsibilities. The implosion program grew quickly, with research activities spilling into other groups within the Ordnance and Theoretical divisions. Men were siphoned off the gun program and assigned to work on implosion. Even with this help, Neddermeyer grew permanently short-staffed as the complicated work unfolded. He pressed his colleagues and Oppenheimer continuously for assistance. Work began to stagnate as the shy man faltered under the pressure and increasingly antagonized Parsons and other group leaders who had their own work loads.

Reports and complaints fell more and more on Oppenheimer. At a November Governing Board meeting, he was compelled to note the "stagnation" and the fact that Neddermeyer's group was too small.[4] Oppenheimer reviewed the history of implosion work for his senior leaders. As far back as April, Neddermeyer had suggested using high explosives grouped in a spherical shell around the active core and tamper assembly. Since then, the Laboratory had learned that small cylinders could be collapsed uniformly when surrounded by explosive charges. John von Neumann had also theorized that the process of collapse became more regular as the ratio of explosive charges to mass went up. Moreover, von Neumann believed the timing involved would be sufficiently quick to support an explosive chain reaction.

Von Neumann's theoretical calculations were invaluable and greatly respected. He was a mathematical genius in the true sense of the word. With a history of major contributions in other laboratories, his work at Los Alamos lent great weight to the implosion movement.

Oppenheimer, like Groves, was keenly aware that implosion had one singular advantage: both uranium *and* plutonium could be used in such a weapon. Groves and Conant were strongly in favor of studying alternatives that would permit the use of both substances. It was still unknown which production process—U 235 or Pu 239—would deliver first and it was possible that Los Alamos would be forced to use one but not the other. Both men also felt that an implosion bomb, if it were successful, would particularly justify Lawrence's massive electromagnetic separation program at Berkeley.[5] These machines employed giant magnets and were costly to build and operate. Groves had to admit that by 1944, Lawrence had produced precious little U 235 with his machines.

The November board meeting discussions reflected the increased interest and concern over the possibilities of an implosion weapon. Work was now underway in a number of Laboratory groups and Neddermeyer's management tasks increased beyond his capabilities with the many small pockets of research. The increased attention of the Governing Board only exacerbated the man's self-consciousness and poor management. Oppenheimer soon received daily reports on the deteriorating program. From Washington, Groves increased pressure on Los Alamos to perfect implosion *and* gun weapons.

The situation grew in the spring to become a crisis, and Oppenheimer was forced to take painful action. For political reasons among his staff, he could ill afford to replace Neddermeyer with an existing staff member. Fortunately, a solution presented itself at the last moment: George Kistiakowsky finally agreed to move permanently to Los Alamos in early summer. Kistiakowsky was one of the few experts on explosives in the nation. Oppenheimer asked him to take charge of the floundering effort and to submit a plan for the reorganization of the implosion project within the Ordnance Division. Privately, Oppenheimer assured his colleagues that Neddermeyer would be moved to a less sensitive position.

On June 13, Kistiakowsky submitted his reorganization plan. Implosion work would now be under his direction as an Associate Division Leader. Neddermeyer would continue as a senior technical advisor, along with Luis Alvarez. A Steering Committee would also be formed with Kistiakowsky as chairman, and with Alvarez, Kenneth Bainbridge, Neddermeyer, and Parsons as members.[6] Oppenheimer approved the plan and undertook the personally distasteful task of informing his friend and former student of the move.

In a tiring meeting on June 15 with Neddermeyer, Kistiakowsky, and others, Oppenheimer reviewed the haphazard organization of the implosion program and its lack of progress. Oppenheimer told the group of Kistiakowsky's plan and his approval of it; Neddermeyer left the discussions in early evening and Oppenheimer prepared a memorandum to his colleague to formalize the new structure.

"The only alternative which has appeared possible to me," he wrote to Neddermeyer, "is to ask Kistiakowsky to undertake the direction of E-5 [Implosion Program] himself. . . . He has been asked to accept full responsibility for the operation of the Group." In a move to soften the blow, Oppenheimer reminded Neddermeyer that he would be a technical advisor attached to the Ordnance Division and a member of the "H. E. [High Explosives] program." He added:

I am asking you to accept the assignment [of new personnel to E-5]. I believe that they are the only ones which have promise of working, and in behalf of the success of the whole project, as well as the peace of mind and effectiveness of the workers in the H. E. program, I am making this request of you. I hope you will be able to accept it.[7]

Oppenheimer realized that he should have acted sooner. The shuffling of personnel revitalized implosion and gave it new status in Los Alamos. The Laboratory settled back into its work. In less than three months, however, the implosion program would again become the center of the Laboratory's greatest crisis, and the basis for Oppenheimer's most drastic reorganization of the Laboratory.

Kistiakowsky's assumption of the implosion program coincided with the arrival of the first plutonium sample from the Clinton reactor. Segré's experiments verified that an impurity in plutonium—the isotope 240—was the source of spontaneous fission, and, if used in a gun-type bomb, would cause predetonation. In order to prevent this, the plutonium gun would have to operate at velocities not possible with conventional muzzles and explosives.

Oppenheimer conveyed the results of Segré's experiments to Conant on July 11. Conant quickly called an emergency meeting in Chicago for the 17th, and asked Thomas, Oppenheimer, Groves, Colonel Nichols, Groves' aide, and Fermi to attend. By evening, they reached the inevitable conclusion: Los Alamos must abandon the plutonium gun and concentrate on implosion.

Oppenheimer and his staff were despondent. Much of a year's work was in vain. The only solution to the predetonation problem in the gun bomb was to find ways of separating out the plutonium 240 isotope. This process would not only require a major investment of resources in a separation plant, but it would also require time to design and build such a plant. A week later, Oppenheimer reluctantly ordered the Laboratory to cease work on the plutonium gun bomb and to shelve plans for further purifying plutonium for the 240 isotope. He asked, however, that steps taken to cease work should not be so irreversible that work could not be resumed at a future date. Oppenheimer was holding out for some discovery that would allow the Laboratory to regain an entire year's work.[8]

For the Laboratory, the first year closed with uncertainty. Oppenheimer had over 2,000 people spread over the mesa and in the canyons of Pajarito Plateau. Suddenly, one major line of work seemed at an inexorable end, with the possibilities for an atomic bomb now considerably reduced. The uranium gun seemed an eventual success, but implosion was quite distant at the moment. There was nothing else to do but throw everything possible behind implosion. For Oppenheimer, the shadows seemed darker: Groves, the President, the millions of dollars spent, the war—all were waiting.

9. LIFE ON THE HILL

There were few roads to Los Alamos. Most new residents took the Taos highway from Santa Fe to Española to catch Highway 4 to Pajarito and the mesa. There was an older, poorly kept road from Pojoaque just off the Taos highway that was seldom used, and even a back road into Los Alamos through the Jemez Mountains that was closed due to snow from November through June.

Almost everyone came first to Santa Fe. Those arriving by train could come only as far as Lamy, a small whistle-stop about 20 miles southwest of Santa Fe. Once there, they would be greeted by old friends or bored GI's dressed in casual clothing. By automobile, the drive was more varied. Many new arrivals from the West Coast saw the vast arid lands of the Southwest for the first time. Those coming from the north or from the eastern states almost always drove through the flat, horizontal midwest or central heartland. Santa Fe was the sleeping jewel at the end of the road.

The small city had barely 20,000 residents and wore comfortably its honor of being the oldest state capital in the United States.

When it was founded in 1609, Don Pedro de Peralta christened it *La Villa Real de la Santa Fe de San Francisco*. Its three hundreds years of history had brought attacks and revolts by Indians, conquest by United States and Army troops, and a final coexistence of Indian, Spanish, and Anglo cultures.

Santa Fe sat languorously at the edge of the Sangre de Cristo foothills and east of the Jemez Mountains. Most of the city was built with adobe bricks that formed low, graceful buildings with a strength and character drawn from the sun-dried mud. The streets were narrow and irregular and often of cobblestone. The center of the city was a public square flanked on one side by the Governor's Palace, built in 1610, and on the other sides by small shops and a Fred Harvey hotel. During the day, the Indians from local pueblos came to sit on the wide sidewalks and sell their pottery and rugs.

Oppenheimer's instructions to newly recruited staff were vague but simple: come to Santa Fe and report to Room 8 at 109 Palace Street. For the first few months in 1943, this address was the new Laboratory's administrative office. Oppenheimer had taken over five rooms in an old adobe building less than a block from the Governor's Palace. Once there, new arrivals walked through an ancient portal with carved wooden lintels called vigas and into a courtyard with an office to the side run by Dorothy McKibben. There was always activity and confusion. As the project grew and the number of arrivals increased every day, the courtyard filled with boxes and suitcases and crying babies. Buses were ordered several times a day from Los Alamos to take the newcomers to the Hill. Everyone received a warm welcome from McKibben, who soothed them and tried to answer their questions. Those with cars were given yellow maps with red markings that led them out of the city and toward Española.

Outside of this small town the ascent to Los Alamos began. The narrow road skirted the San Ildefonso and Santa Clara pueblos and crossed the Rio Grande River at Otowi on a frail wooden bridge to begin slowly climbing into the Jemez Mountains. At Otowi, not far from the river crossing, lived Edith Warner and her Indian companion, Tilano. Both of them were in their late fifties, and together they ran a small tearoom that had been a stop for the narrow-gauge Denver & Rio Grande railroad, nicknamed the Chili Line. Their tearoom would soon become a sanctuary for tired scientists and their wives.

Highway 4 was as formidable to travel as it looked. The Los Alamos Boys' School had never bothered to do anything more than have the road cleared and leveled every few years. For most of the first spring, Oppenheimer and his scientists found it better to make the trip from

Española on a tortuous, unpaved road strewn with rocks that ranged from pebbles to boulders. As the ascent sharpened, the road began to bend into sharp switchbacks, with edges that seemed to crumble dangerously with every vehicle that passed. The rugged beauty of the terrain was quickly lost on travelers as they became increasingly concerned for their lives. Even Groves was appalled and ordered the road paved. Even that was not enough, for the steady travel by heavy trucks and cars seemed to tear the road apart after only a few weeks; the larger trucks could barely navigate the switchbacks. Some turns were so sharp that larger trucks had to back up at the switchbacks in order to make it up to the mesa. Groves was again forced to undertake a major project by widening and strengthening the road.

Surviving the journey, however, was only half the experience. Just as the road leveled and straightened, a tall wooden watchtower loomed into view. Near its base was a small house guarding the entrance to Los Alamos and staffed with fierce-looking military police. It was to be everyone's introduction to the Laboratory's security system.

New arrivals were asked to leave their cars or buses and undergo a security check. Names were studiously checked against rosters and preliminary instructions were given on how to drive, where to go, and what not to do. Temporary passes were issued to each adult member of the family until permanent cards could be prepared for everyone in a day or two. Slightly shaken, the arrivals were told to go two or three miles, through another checkpoint, and finally onto the mesa proper. At last, the bewildered arrivals had their first, unobstructed view of Los Alamos.

What rustic beauty the Boys' School had preserved and nurtured had been ravaged by the Army. The lovely log buildings of the school were now hidden by a melange of ugly prefabricated green-and-gray structures. The road was dirt and seemed to lead without purpose through the array of one- and two-story buildings. Only the tall, darkened Jemez Mountains seemed to break the ugliness with a reminder of what the mesa was once like.

Even worse was the realization that one would have to live and raise children in this setting. Dazed, everyone proceeded to a housing "office" in a garage converted for the purpose. With housing assignments in hand, the new arrivals would often have to meander through several residential areas that looked like mining camps in order to find their own apartment complex. Each new employee and his family had been prepared somewhat for the inevitable trauma by memos labeled "restricted" and mailed to each new staff member as a preparatory introduction to life on the Hill.

These memos were titled *Memorandum on the Los Alamos*

Project and were issued in a series: First Memorandum, Second, and so on. The first memo had been prepared under the watchful eye of General Groves and reflected his concern with security. As little information as possible had been given. With growth and complaints, however, the Laboratory found it necessary to prepare succeeding arrivals with more "background information." The memorandum began ominously enough with a warning that the reader was to share the document with no one not known to a member of the "Project," and with his wife, *only* if she agreed to silence.

With that, and with a citation of the Espionage Act, the memo introduced the reader to Los Alamos. It was situated, they were told, at 7,300 feet and on a strip of land two miles wide by eight miles long. Cautiously, it added that most of the construction was by the U.S. Army. Schools, commissaries, beer halls, and theaters would be available, as would community services, such as water, electricity, and sewerage. The weather was purported to be pleasant—an average of 67.7°F—and rarely hotter than 95°F in the summer and −14°F in the winter. As a precaution, there was a note to supplement ordinary clothing with "rough country clothes."

Robert and Bernice Brode were one couple who arrived with their two sons in September 1943. They presented themselves at 109 Palace Street under a sign that said "U.S. Army Corps of Engineers." Met by Dorothy McKibben, they were directed up to Los Alamos, where they were assigned Apartment C in Building T-124. There were no street signs, but directions were given in relation to the Laboratory's wooden water tower. After some confusion, they found their fourplex on the far end of a slope looking into the Jemez Mountains. They were quickly greeted by the wives and children of the Bacher, Cyril Smith, and Teller families. There was little time for formalities, and with introductions over, everyone—new and old—settled into the peculiar life-style of Los Alamos.[1]

Most families were concerned first about housing. Single men and women were shunted off to apartment complexes—organized by sex—where they were given single rooms separated by a bath. Married couples without children received one-bedroom apartments, usually in a complex of four or six units. Couples with children were given two- or three-bedroom apartments, depending upon the number of children, their prominence as scientists, and most importantly, the apartments' availability. The scramble to construct housing was a continuous headache for Oppenheimer and Groves, who, from the moment Los Alamos was chosen as a site, persisted in underestimating the number of scientists and technicians needed. The shortage continued until well past the end of the war.

The original Boys' School complex included a row of small, beautifully built cabins and homes that were used for faculty and student activities. One of these, not far from Fuller Lodge, was remodeled and given to Oppenheimer's family that included a son, Peter, and daughter, Toni. The few others were doled out to senior Laboratory leaders, such as Captain Parsons. For a while only this row of homes behind Fuller Lodge possessed bathtubs. Everyone else in Los Alamos had a shower. When this was discovered, Oppenheimer's neighborhood was nicknamed Bathtub Row.

Oppenheimer, Manley, and other early planners had not totally rejected aesthetics. Left to the Army, the housing areas would have been built without concern for the terrain and arranged in long, neat rows. Oppenheimer had argued that the natural beauty of the mesa could be taken advantage of by following the ground contours. As families arrived, additional pressure was put on both Oppenheimer and Groves by Los Alamos wives to be more mindful of the mesa's natural beauty. While accommodations were made, and while housing areas still looked sporadically built, they were often arranged in semicircles and around irregular ground. Some effort was made to protect many of the tall, graceful trees that covered the mesa.

Much of the housing took on the name of its builder. One of the earliest contractors was the Sundt Company of Santa Fe. They constructed a large number of shoebox buildings that contained efficiencies and two- and three-bedroom apartments. Many of these units became status symbols as Los Alamos grew and more housing was built in remote areas of the mesa. Some of the more prominent Sundt residents included Fermi, Teller, and Bethe. The Morgan Company followed Sundt and built a large number of duplexes. Still later, the McKee Company came and built a series of flat-roofed single-family homes. Another prefabricated type of home was built toward the end of the war and called the Pasco home. It was almost entirely square, like a box, and was modeled after the sterile homes built by the Manhattan Project at its Hanford Plant in Washington. Most of the early Sundt apartments tried to incorporate Oppenheimer's pleas for amenities. The floors were hardwood and varnished,and and many contained fireplaces. While only the unmarried staff apartments were furnished, each apartment or home was provided with a new refrigerator and a wood and coal cooking stove.

There were at least two bright spots to life at Los Alamos: rent and medical care. Unmarried men and women paid no more than $13 a month for rent and families paid rent based on the husband's income. Someone making less than $2,600 a year paid only $17 a month in rent; those making over $6,000 paid no more than $67. A barracks-style hospital had been built and was staffed with two full-time doctors and several nurses. Medical care was free and provided on a first-come basis. Hospital stays cost a dollar a day.

After a family was ensconced in housing of one sort or another, its second hurdle was acclimation. The heavy cast-iron stoves were slow to start and produced a great deal of heat during use. In the summer months the kitchen was like a steam bath. The apartments were for the most part heated by coal, which generated clouds of soot. In the first cold days of fall, an enlisted man would start the heater in each home and apartment complex and a fine layer of black soot was sprayed over everything near the vents. Each roof sported a chimney that contributed the same fine covering to surrounding surfaces.

Most women had been accustomed to modern conveniences, such as gas cooking stoves. The "Black Beauties" they found waiting for them were always a shock, and almost everyone had trouble starting the wood ranges. Robert Wilson's wife, Jane, struck a blow for all wives one evening when she cajoled General Groves into a demonstration of igniting the wood-burning stoves. After an hour and a sooty uniform, Groves produced a small flame. The next morning, Army-issue hot plates were given to each apartment in Los Alamos.

The streets were unpaved and had no names. Apartment units all bore military designations, and visitors or new arrivals could judge direction only in relation to the water tower near the commissary. When it rained the streets turned to mud, and cars were often abandoned until the weather cleared and the streets could dry out. When it didn't rain, there were shortages of water. In fact, despite the addition of one water line after another, Los Alamos seemed permanently settled in a drought.

There were cycles of famine and plenty in food items like fresh vegetables and meat. The usual facilities, such as the laundry, were convenient only if you lived nearby. Many wives, accustomed to life in larger cities, found themselves without domestic help. Maids were available but were limited to the local Spanish or Indian women from the villages below Los Alamos.

Each day Army buses were dispatched to nearby pueblos and to Spanish-American settlements to pick up those women who were willing to work on the Hill. They were brought to the Maid Service Bureau located in the garage of the old Boys' School. Inside one of the two rooms a list would be posted, indicating which home or dormitory required service. Each woman would be given morning and afternoon assignments. No one in Los Alamos was entitled to more than a half-day of maid service. At the end of the day, the women would return to the Bureau office or to the PX to wait for buses back to the valley.

While work provided these women with money never before earned, they observed a series of religious feast days, as well as days of lack of interest. Many wives could hardly understand the cultural forces at work. There were inquiries and anger—with complaints, of course, made to Oppenheimer—and in the end Los Alamos had to settle for what it could get.

Schooling for Los Alamos children was another problem. It had not escaped Oppenheimer's planning that a school should be built for Hill residents with school-age children. Indeed, contrary to most Los Alamos housing, the school building was a model of sound construction and relative beauty. Built of cinder block, the building was spacious and faced the Jemez Mountains on one side through glass windows. In the evenings, the building was intended as a community center. At the last minute, it was realized that the "young" quality of the mesa—the average age of the adults was the middle twenties—also meant many young children of preschool age. Oppenheimer and his contractors had to hurry to build a nursery school for its under-six population.

One of the first employees of the new school was an industrial arts teacher whose original orders for materials consisted of an array of mechanical equipment. In a letter to General Groves, Oppenheimer was forced to expand the scholastic program of the school and to point out that the industrial arts teacher had been hired by the Post's military command. The need for quality educators, he added, was hardly a "frill" when the general education level of the populace was considered. Despite the high level of education, however, Oppenheimer conceded that the new school would allow the children of the mesa's Spanish population to attend.[2] Very quickly, the Hill created its own school board composed of Laboratory residents.

General Groves seemed unconcerned. His original concept of the Laboratory was a military post without wives and children. Pressured by Oppenheimer and others into accepting a "family" setting, he seemed to resist understanding basic human nature. When informed of the rising birth rate in Los Alamos, Groves was heard to remark that "it had to stop." Hill residents firmly believed that Groves was behind the slim stock of baby foods and diapers in the commissary.

In spite of difficulties, life on the Hill took shape. Wives met startled newcomers to the neighborhood with smiles and coffee. A network of baby-sitters emerged and prospered. Agreements and moratoriums were negotiated. Pianos and radios were played during "open" hours by agreement among tenants of the same building and a woman baby-sitting one day could expect the favor to be returned. Trips to nearby Española and Santa Fe were combined to use only one car and fewer gas coupons. Everyone tried to remember which wife and husband belonged to each other. Oppenheimer came to be cursed less and less.

Even the physical ugliness of the town was revitalized by the energy of community groups and most of all by the majestic scenery around Los Alamos. From one flank rose the Jemez Mountains, and from the other the Sangre de Cristo. Almost everyone fell in love with New Mexico and Pajarito.

Life in Los Alamos was singular enough because of its physical setting. But there was another factor that permeated everyone's life: security. Oppenheimer was never far from Groves and the persistent demand for tight security. In early 1943, the battle for civilian control of Los Alamos had barely been won. Groves had reluctantly agreed to hold off mobilizing scientists until June, 1944—a year away. Oppenheimer, however, agreed in principle with Groves on the need for security at the Laboratory. It was necessary both to warn prospective staff about security as well as to downplay its pervasiveness.

In the beginning, Oppenheimer, Manley, and others recruited staff for Los Alamos by combing universities and scientific projects. Both teaching staff and graduate students possessed a good deal of personal freedom. Life in Los Alamos would call for stringent restrictions on personal mobility, as well as on communications with friends, family, and colleagues.

Oppenheimer tried to prepare people in his interviews and instructed other recruiters to do the same. Oppenheimer went so far as to prepare a "Note on Security" for the memoranda sent to individuals before they moved to Los Alamos. Most scientists had been shown the letter from Groves and Conant to Oppenheimer citing the important wartime nature of the project and emphasizing the need for secrecy. Oppenheimer cited the Groves-Conant letter and optimistically stated that "travel" to and from the Los Alamos Post would not be restricted to permanent members of the staff. Travel outside the mesa, however, did not mean unlimited travel, but rather movement that related to Laboratory work or perhaps to personal emergencies. Oppenheimer's most severe note came in the last paragraph:

The extent to which we shall be able to maintain this comparative freedom will depend primarily on our success in keeping the affairs of the Laboratory strictly within the confines of the Laboratory, on the cooperation which the project personnel affords us in its discretion on all project matters, and on our willingness to rupture completely our normal social associations with those not on the project.[3]

While there were objections to this, Oppenheimer was powerless to do otherwise. Men still came and, despite the preparation, they and their families were still shocked and disconcerted to discover how truly restricted their lives were to become.

Not only would they be confined to Los Alamos, with brief excursions to Santa Fe the exception, but they were not permitted to have visitors or friends on the Hill. Everyone was given the same address in Santa Fe for correspondence: P. O. Box 1663. Only close family members could be told that the project was in New Mexico. Professional magazines and journals could not be received directly but were required to be forwarded through former university or job offices. If that was not possible, the individual could use a blind post office box in Los Angeles. Groves hoped that this circumspection would prevent enemy agents from compiling lists of personnel at Los Alamos.

Eventually, even personal mail was subjected to scrutiny. Censorship of the mail did not begin until late in 1943. It had not occurred to Groves and Oppenheimer that mail needed perusal until rumors began to circulate that all mail was secretly being opened and read by Army officials. Complaints were filed with Oppenheimer, who in turned protested to Groves. An investigation initiated by Groves revealed no censorship but raised the question of need. Groves decided that censorship was needed and persuaded Oppenheimer. At a Governing Board meeting late in October 1943, Oppenheimer asked his senior staff if they had any objection to censorship of the mail and, if not, did they feel they were speaking for the Laboratory as a whole. No one objected and there was a consensus that while censorship would be an additional hardship, the Laboratory would accept it.[4] Oppenheimer notified Groves, who initiated a laborious process of reading all mail in the cramped quarters of the Santa Fe offices on Palace Street.*

A list of regulations was prepared by the Army and circulated to each staff member and his family. Scientists were forbidden to enclose personal mail with official mail to other Manhattan laboratories. Private mail was to be placed in unsealed envelopes and dropped in special mailboxes scattered around Los Alamos. Sealed mail would automatically be returned to the individual by the censors. Mail that contained objectionable content would not be excised or obliterated

---

*Censorship was supervised by Army Captain Peer de Silva, who in the early 1950's would figure prominently in Oppenheimer's security hearing.

but simply returned to the sender with notations on the offending items. Incoming mail, however, would be opened and read and resealed with official censorship stamps and seals.

Oppenheimer tried to avert staff reactions by presenting the new regulations in a preventive light. The intention, he said, was to prevent the Los Alamos project from being connected with other projects in the United States and to prevent details of its size, physical characteristics, and the identity and numbers of personnel from falling into the wrong hands. It was emphasized that while many details were insignificant by themselves, they could, if pieced together, form some picture of Los Alamos and generate interest by the enemy in further spying and sabotage.

More specifically, Los Alamos citizens were forbidden to reveal their present location. In personal emergencies, they were permitted to say they lived in New Mexico. Nor were they to reveal the names of their associates at the Laboratory. There could be no references to the professions of personnel, such as physicist or chemist and, of course, no mention of one's own work or the equipment used. Although it struck many as amusing, if not ironic, the rules even forbade the mention or photographs of housing at Los Alamos.

Security had its lighter side as well. Letters sent to the commercial firms were occasionally returned to Army censors with a note that the sender failed to enclose a check. Letters with drawings or doodles were also returned. The wife of Richard Feynman, a young physicist working with Teller, was in a hospital in Santa Fe with a serious illness. To amuse her, Feynman devised simple codes and wrote her cryptic letters every day. The censors quickly reacted. Major Peer de Silva, the Post Security Officer, called on Feynman to administer a grilling. Feynman explained the purpose of the codes and offered to supply de Silva with a deciphering table; de Silva was not amused and ordered Feynman to stop.

The Post Security Office also arranged for drivers' licenses to be made without addresses and only a code number. More than a few irate policemen found their offenders with foreign names, and no address or occupation to admit to. Major figures were assigned code names to avoid attention. Enrico Fermi became "Henry Farmer," and Niels Bohr became "Nicholas Baker." Everyone became an "engineer," and if pressed by local policemen or officials, was told to say he worked for the Army Engineering Corps. Secret numbers were given to automobile registrations, bank accounts, income tax returns, food and gasoline rations, and even insurance policies.

The concern for security was particularly strenuous for key

members of the Manhattan Project. As early as 1943, General Groves had forbidden Oppenheimer and Ernest Lawrence to travel on airplanes. It wasn't until the war ended that Groves withdrew his restrictions. Security "shadows" were assigned to Oppenheimer when and wherever he traveled. In Los Alamos, Groves ordered special MP's to guard the homes of Oppenheimer and Captain Parsons. This was particularly hard on the wives and families, who would occasionally forget their identification badges and would be forbidden entry to their own homes until proof could be produced.

No one really knew what Los Alamos would be like when he arrived, and almost everyone was surprised at what he found. A few believed the security too pervasive and felt as if they were always being watched. Others wondered whether the high fences were to keep intruders out or to keep the residents in. Almost everyone adjusted as best he could. For the men, the work was fascinating and meant a six-day week; for the families, the days varied in their pleasures and toils. Oppenheimer tried to be a model. There was little to be done, he said, except to make the best of it.

The work day in Los Alamos began at 7:30 A.M. with a shrill siren that sounded from the Technical Area. Another siren would sound 30 minutes later. As the war progressed the sirens were sounded more often in the early morning. Except for the hottest days in summer, and days of rain or snow, the morning was always crisp and the mountain air electric. When he was in Los Alamos, Oppenheimer could be seen walking quickly to the Administrative Building, and was among the first to his office. The days were long, and work drifted into the evenings as well. Sunday was an official day off, but as 1945 neared, the work week was almost always seven days.

Husbands would be fed and packed off to work and soon thereafter the children would be sent to school. Wives would then be free for more coffee or to rush to the commissary or do laundry before the others. The community's washing machines rented for thirty cents an hour and mangles for forty cents. It took a while for the Army to accustom itself to serving what it no doubt regarded as a soft constituency. The first year, many of the commissary's shelves were lined with gallon tins of tomatoes or corn, or huge bottles of imitation flavoring, such as vanilla or lime. There could be curious exceptions that would appear once and never again: bottles of stuffed olives, chocolate, and brandied figs. Perishable items such as meat and vegetables were ordered from Texas, until residents pointed out to Army officials that better vegetables and meat could be had locally in New Mexican valleys. Local Indian girls were originally hired to run the cash registers until the Army discovered the civilians cajoling the young girls out of more than ration stamps allowed. WAC's were finally imported to run the commissary.

The Post Commander created a small daily newspaper called the *Bulletin*. At first it served as an official organ of Army news. Later, the newspaper mellowed and local news included births, babysitting services, items for sale or wanted, and dozens of local minutiae that were the substance of daily life. The *Bulletin* also warned residents of wrong behavior or impending disasters. There was an almost constant crusade to persuade families to cut down on the use of water. One *Bulletin* in 1945 carried an extensive list of what to do to conserve water: don't wash automobiles or water lawns; turn off water while soaping in the shower; and don't let the water run to cool. Young GI's raced around the mesa in the early morning to stick copies in door handles or back porches.

The *Bulletin* was not a substitute for magazines or newspapers. One small log cabin built by the Boys' School was converted into a PX where people could buy magazines, newspapers, cigarettes, and soda-fountain treats. The cabin quickly grew too small, and the Army announced plans to level it and build a new PX. There was a great outcry from citizens, but to no avail; the town simply awoke one day to find a new PX under construction.

Some wives worked for the Laboratory. Many of the women were in secretarial jobs, but others were in technical positions or served as human "computers" of long mathematical problems. Everyone who worked for the Laboratory had a special color-coded identification badge which admitted him or her to various buildings or areas of the Laboratory. White badges were the most encompassing and the most prestigious: their possession meant access to the entire Technical Area or to any of the remote locations such as Anchor or S Site. Other colors meant access to only specific buildings or locations.

All wives kept busy. Kay Manley directed the town's choral society. Eleanor Jette and Peg Bainbridge took on the Cub Scout Pack, and Dorothy Hillhouse ran the Brownie Troop. Bernice Brode, like dozens of other women, worked part time in the Tech Area as a statistician. Both Jean Bacher and Eleanor Jette served on a Housing Committee for the fledgling Town Council. Several wives taught in the Los Alamos schools. Alice Smith taught social science, Betty Inglis taught mathematics, and Peg Bainbridge taught French. And Kitty Oppenheimer acted as an unofficial hostess for Los Alamos.

In the beginning the Town Council was appointed by Oppenheimer; later it was elected. Although it had no real power, it was able to discuss problems related to the community and make recommendations to Oppenheimer or the Army Post Commander. Agendas ran from planning festivities to investigating the charge that prostitutes were propositioning young men outside the PX.

Life on the Hill included parties and weekend fun. The parties were very often the highlight of the week and were, as veterans recall them, unmatched for gaiety and alcohol. Dances were frequent: square dancing, particularly, was popular and was taken up as a local sport. Always there was lots of beer. Singing groups were organized, as well as drama groups, and after a few months a small symphony orchestra was assembled.

Many parties were given by the single men and women who lived in the dormitories. These were the loudest, most crowded, and most alcoholic on the Hill. Each dorm had a large lounge, which was emptied of furniture and arranged to accommodate a small band and tables for food and drinks. The parties lasted until dawn, and when couples tired of the noise or dancing, they drifted into the rooms for quieter socializing. The focus of the party was the punch bowl, usually a huge five-foot glass chemical reagent jar from one of the nearby laboratories. The punch itself was usually a mixture of whatever liquor could be purchased in Santa Fe for the occasion.

One party was particularly extravagant. From the British Mission in Washington, D.C., Lady Chadwick, wife of Sir James Chadwick, authorized the British contingent at Los Alamos to host a party. The war was nearly over, and Lady Chadwick decided that a farewell party was in order. Special invitations were printed and placed into the mail dockets of the Tech Area. The dress was an exception to the informal style of Los Alamos: men wore at least a suit, and a brave few sported tuxedos. Many women attended in long dresses with gloves.

Rudy and Genia Peierls were the official host and hostess. Good liquor was ordered from Santa Fe, and wives from the British Mission spent most of a day preparing "English" food for their guests. No one had enough wine glasses for the occasion and the local Woolworth's in Santa Fe was persuaded to rent dozens of glasses for the party. A formal high table was set for senior members of the Mission, as well as Laboratory leaders like Oppenheimer. Several wines and liquors were served. The evening produced a series of skits spoofing Los Alamos life, and even included a dramatization of Trinity with the tower and Fat Man represented by a ladder and bucket.

Other parties ran from conventional to the elegant. There were simple meals for friends and neighbors or dinners with silver service and finger bowls. The Laboratory's large foreign population made Italian and Hungarian food popular. Larger and more sober parties could be held at Fuller Lodge for a rental of $5 a night. Divisions in the Laboratory rented the Lodge and held parties for their staffs and families. The Lodge also served as a theater for dramatic productions. One Los Alamos play was a spoof that included characterizations of General Groves and Oppenheimer. Another performance included a bit part played by Oppenheimer, who made his entrance onto the stage in a coffin.

The weekends also provided opportunities for Hill residents to discover New Mexico. Los Alamos sat in the middle of striking landscapes and ancient cultures. Fermi was a year-round hiker, and often organized scientists into outdoor walking and working sessions. Los Alamos was on the eastern edge of the Santa Fe National Forest and driving or hiking west into the Jemez Mountains meant places to visit or picnic, and perhaps the chance to discover some new nook or portion of a lonely canyon or cliffside.

Most residents did not find it necessary to travel far. Pajarito Plateau was divided again and again into tortuous canyons. The technical and residential areas of the Laboratory were almost entirely located on Los Alamos mesa. To the south and north of the mesa were Los Alamos and Pueblo canyons, and with a casual hike or drive, families could wander further north and find themselves in Mortandad or Sandia canyons. There were numerous creeks feeding into the Rio Grande that would appear and disappear between rain and dry weather. In spring, winter snows melted and filled the dry creek beds and the season nourished the rabbit brush and shrubs called Apache plumes.

The canyons and cliffs wore a visual history of the birth and aging of the land on their surfaces: volcanic thrusts, cliffs worn and weathered from age and storms, and canyons gorged and carved through endless seasons. Each fall and spring accentuated the landscape with changes of color and texture; even winter could sometimes bring out a stark beauty to the harshness of the land.

Scattered throughout the plateau were abandoned pueblos and Indian ruins. South of Los Alamos, in Frijoles Canyon, were the ruins of Bandelier. Frijoles was a great gash in the pleateau that ran nearly 17 miles in length and in places was over 600 feet deep. All along the cliffs were caves carved into the soft pumice walls by Indians in the thirteenth century. Even the floor of the canyon contained the remains of large community buildings and ceremonial kivas, or meeting places dug into the ground. Only a few miles away were the ruins of Tsankawi, which the Tewa Indians had called Place of the Round Cactus. Highway

4, as it cut through Pajarito and yellow pine and juniper trees, revealed the distant canyons honeycombed with cliff caves. East of Los Alamos, set on a rise in the canyon floor, were the ruins of Otowi, and farther north was the ruined cliff city of Puye. On the right kind of day, and with the wind in the trees, one could imagine Indian families moving among ruined walls or scrambling up wooden ladders for safety in the cliffs.

In the valleys below Los Alamos, nestled against the Rio Grande, lived the descendants of the ancient Indian tribes. San Ildefonso was one of the largest pueblos, and became a source of labor for the Laboratory. One of its most famous residents was Maria Martinez, whose careful cultivation of Indian pottery techniques revived the art form from near-extinction. San Ildefonso, like other pueblos, had a curious mixture of Indian and Spanish cultures, epitomized in the coexistence of pagan and Christian religious beliefs and ceremonies.

The Spanish had come to the valley in 1598 and built a church at San Ildefonso some years later. The Indians adopted Christianity, but continued to revere a nearby vacant mesa called Tunyon, or Black Mesa, as the home of their ancestral gods.

Not far from San Ildefonso were the pueblos of Santa Clara, Nambe, Santo Domingo, Zia, and San Juan. And sixty miles north was the multistory pueblo of Taos, whose long-used adobe rooms made it the oldest continuously inhabited dwelling in North America. The plateau seemed juxtaposed between new and old, and in the midst of it all was the new Laboratory trying to unlock age-old secrets.

Below the Laboratory, sharing the valley with the Indians, was Edith Warner's tearoom, a place reserved by Hill residents for special evenings. This remarkable woman had come to New Mexico in the 1920's for her health, and stayed. She was joined a few years later by the oldest Indian at San Ildefonso. Together, she and Tilano managed the Otowi Station of the small railroad that ran near the Rio Grande. Her gentleness with the nearby Indian families and her love of the land earned her the devotion of the Pueblos. She was permitted to live on the San Ildefonso reservation and had an adobe house built for her. Through most of the 1920's and 1930's her small home was a stop for tourists, whom she fed with home cooking and vegetables and fruits grown in her own garden. Oppenheimer had discovered her house in 1933 while on a summer holiday at his ranch. The two spent all afternoon together and became good friends.

With the creation of the Laboratory, Oppenheimer began to take his colleagues to Edith's tearoom for a quiet evening and dinner. The Bachers, McMillans, Bethes, and Niels Bohr all began to go regularly and take others. Eventually, Oppenheimer asked Edith to serve only Los Alamos residents and to close her small restaurant to outsiders. He could not tell her why, except to say that their work was very important to the war and very secret. Edith Warner asked no questions, and prepared dinner every day for ten people. She never accepted more than two dollars from each person, and was never known to make a profit.

The evening was always memorable. The tearoom was actually a kitchen and two small dining rooms with an adobe fireplace in each corner. Miss Warner always cooked on an old black stove that burned wood, and her kitchen smelled of freshly prepared bread and vegetables. There was no electricity, and the only light was provided by the fireplace and the soft glow of candles on each table. Tilano chose only piñon wood to burn quietly and to scent each room softly with its fragrance. The food and atmosphere made her famous among Hill residents and she was soon beseiged with requests for reservations weeks ahead. Eventually, she was forced to close her tearoom except for Robert and Kitty Oppenheimer, who would come down to dinner once a week. Hill residents drifted down during the day, and on weekends several helped to build her a new house a few miles away.

The Indians took the burgeoning Los Alamos population in stride. For some, Los Alamos meant jobs as construction workers or maids; for others, the newcomers added to the market for pottery and crafts. They seemed to the Indians to be no brasher than the general lot of tourists. For the Hill, however, the Indians were more interesting. They were a source of cheap but unpredictable labor, but mostly they were seen as colorful and quixotic. A few Hill residents became serious students of Indian life and history, but most regarded the Indians pleasantly or condescendingly and photographed and catalogued them as local attractions.

The reaction in Santa Fe was somewhat different. Residents of that ancient city were at first amused with the new arrivals and considered them, as did the Indians, as merely tourists. Their voices and baggage and unruly children seemed only to confirm suspicions about the decadent East and West coasts. As Los Alamos grew, however, so did the demand on Santa Fe for gasoline, food, home items, and the like. The money Hill people spent barely offset the invasion of Laboratory buses that congested the archaic streets and the speeders and other wrongdoers that always seemed above the law. The U.S. Army seemed particularly disposed toward intervention.

For Los Alamos, Santa Fe was an oasis of relief and frustration. As individuals governed by the pervasive security system, they were angered to learn that Army security men posed as bartenders at the

La Fonda Hotel, waiters in other restaurants, and salespersons in countless stores—or so the Hill people thought. The same security measures prevented Los Alamos residents from mixing with Santa Fe's interesting and eccentric artist and author conclaves. All of the beautiful homes were closed to Hill residents, except for occasional glimpses through an open door or gate. Ultimately, Santa Feans were tolerant. The city's age and cultural confluence gave them a snobbery that was curious but not inquisitive. Visitors from Los Alamos always came and left, usually in one day, and returned to security guards, water shortages, and dusty streets.

In Santa Fe, it was rumored that Los Alamos was assembling the rear end of horses for the U.S. Government. Los Alamos made no reply.

Life on the Hill /107

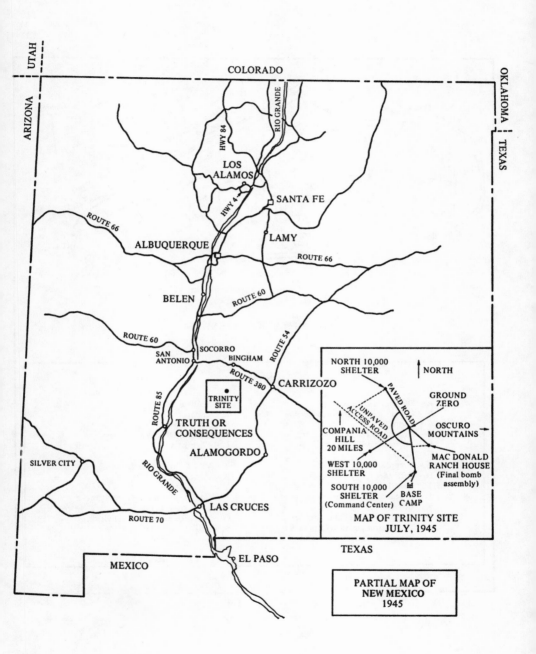

108/ Alpha

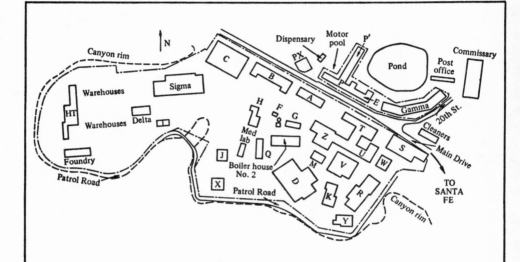

MAP OF TECHNICAL AREA – 1945

LEGEND

A - Director's Offices
B - Offices and Conference Rooms
C - Shops
D - Plutonium Purification Plant
E - Theoretical Division Offices
F - Storage
G - Graphite Fabrication
H - Offices & Laboratories
HT - Heat Treatment Plant
J - Research Laboratories
K - Gas Stock
M - Chemistry Supplies & Storage
P - Theoretical Division Offices & Personnel Offices
Q - Medical Offices
R - Laboratories
S - Stockroom
T - General Offices & Library
U - Chemistry & Physics Laboratories
V - Shops
W - Van de Graaff Machines
X - Cyclotron
Y - Cryogenics Laboratory
Z - Cockcroft-Walton Generator
Delta - Auditorium
Gamma - Research for M Division
Sigma - Metal, Plastic, & Ceramic Fabrication

10. THE SECOND YEAR, 1944-1945, PART I

The elimination of the plutonium gun left the Laboratory with only the implosion bomb and the uranium gun. Oppenheimer acted quickly to formalize a program for implosion within an administrative structure organized to produce "guns." In Europe, the war continued to shift in favor of the Allies, although there was still the possibility of a German atomic bomb. The American war arsenal could only expect use of the bomb in Japan; more than likely it would be a plutonium bomb, for a bountiful supply of uranium looked far away to eager Los Alamos eyes.

Oppenheimer had already been constrained to explain to the Governing Board that it was difficult and costly to have the Laboratory continue simultaneously with both gun and implosion work. With the results of Bethe's tests, he said, it would be possible to know which direction to follow. In May 1944, Conant met with Oppenheimer in Chicago and urged him to pursue an implosion weapon even if it were to be a low-efficiency one. Conant argued that both U 235 and Pu 239 should be considered for such a weapon. Even a bomb with a force of several hundred tons of TNT would be acceptable.[1] On July 20, 1944,

Oppenheimer ordered that all possible priority be given to the implosion program without restraining completion of the uranium gun. Both Oppenheimer and Groves felt the time now warranted a major reorganization within the Laboratory.

Two new divisions were created by Oppenheimer, the G, or Weapons Physics Division, and the X, or Explosives Division. Weapons was headed by Robert Bacher, who had shown strong leadership during the last year. The new division incorporated parts of the old Experimental Physics Division, as well as several small groups from Ordnance. George Kistiakowsky took leadership of X Division. Experimental Physics was renamed R, for Research, and given to Robert Wilson, and Parsons continued with his remaining Ordnance staff.

George Kistiakowsky was a fortunate catch for Oppenheimer. The Russian-born chemist was one of America's few explosive experts, and his work in the field of chemistry had earned him a strong reputation at Harvard. Kistiakowsky was the chief of the National Defense Research Council's Explosive Division when Oppenheimer asked him to come to Los Alamos. At first, Kistiakowsky was reluctant to leave the NDRC. Conant joined Oppenheimer in urging him to join the Laboratory staff. He agreed, and came in time to salvage Neddermeyer and the implosion program from stagnation. His name was quickly reduced to a more manageable form: Kisty.

Enrico Fermi finally disengaged himself from his work at Chicago and was able to come to Los Alamos and join the Laboratory staff full time. Fermi and Parsons became associate directors of the Laboratory, Fermi with responsibility for F Division which included research, theory, and nuclear physics, and Parsons, in charge of ordnance, assembly, delivery, and engineering.

Work on both the gun and implosion weapons now required closer coordination between several teams working on different aspects or components of the same weapon. The Weapons, Explosives, and Ordnance divisions were to engage in closely interrelated work that required each to share facilities, equipment, and, occasionally, personnel. It was imperative that all plans and component specifications of these divisions be incorporated into the final weapon designs. For the implosion bomb, the Theoretical, Chemistry and Metallurgy, and Research divisions were in a support capacity.

Oppenheimer was overburdened with technical and administrative work. The divisions of the Laboratory had been created in an atmosphere in which the state of weapons technology was considerably less well defined. Research and development the first year had made necessary a number of *ad hoc* decisions and temporary mobilizations of personnel that had somehow continued as permanent units. Although Oppenheimer shared responsibilities with his Governing Board, it, too, had become seriously overburdened. Members of the board not only kept their responsibilities for research and development, but also took on numerous administrative tasks, such as mediating housing disputes, contending with complaints about water and food shortages, drafting policy for the Laboratory, and generally acting as funnels for complaints to Oppenheimer.

Oppenheimer took two additional steps. A third division was created for administration, and the Governing Board was dissolved and reorganized into two smaller boards, Administrative and Technical. Oppenheimer carefully laid his groundwork with both the Laboratory leadership and General Groves. On June 29, he formally announced the dissolution of the Governing Board and gave his reasons: the burden of the board had increased tremendously since its creation and there was a need to handle technical and administrative issues separately. His action, he said, was supplemented by a need for planning for future requirements.[2]

The Administrative Board was organized informally, and members were encouraged to raise any questions concerning administrative problems. The board could extend invitations to other members of the Laboratory to discuss specific topics. Oppenheimer asked the Post Commander, Lieutenant Colonel Whitney Ashbridge, to join, as well as Bacher, Bethe, Dow, Kennedy, Kistiakowsky, Mitchell, Parsons, and Shane. Ashbridge's membership on the board was an example of Oppenheimer's continuing attempts to involve the Post's military personnel in the workings of the Laboratory. Ashbridge was an intelligent man and no newcomer to Los Alamos. As a youth, he had attended the Los Alamos Boys' School.

The Technical Board was charged with considering immediate technical concerns and was to review and assess the progress of work within the Laboratory. The board consisted of Luis Alvarez, Bacher, Bainbridge, Bethe, Chadwick, Fermi, Kennedy, Kistiakowsky, McMillan, Neddermeyer, Parsons, Isidor Rabi, Ramsey, Cyril Smith, Teller, and Wilson.

The new Administrative Division became a hodgepodge of individual offices formally under Oppenheimer, and included Personnel, Business, Procurement, Library, Health, Patents, Safety, and now the Shops and Maintenance. David Dow became Oppenheimer's assistant and handled administrative matters of a nontechnical nature and acted as continuing liaison with the Post Military. Into Dow's hands fell the perpetual and seemingly insolvable problems of water shortages, insurance, housing, and salary inequities.

Almost immediately the new Technical Board began to falter as implosion work gathered momentum and need for tight scheduling increased. The board was never formally dissolved, but Oppenheimer increasingly gave responsibility to special committees and task forces. This shift from a single administrative board to more specialized committees was no doubt intentional by Oppenheimer. It permitted him to draw on individual or group talents from all over the Laboratory and place them into critically needed areas. It was a return to a more informal but more flexible structure, and marked Oppenheimer's consolidation of authority.

Oppenheimer was also able to use individuals in "floating" consultant roles without offense to established leadership. In 1944, Sam Allison, the director of the Metallurgical Laboratory, arrived in Los Alamos as a consultant. Within a few months others arrived, including Niels Bohr and his son, Isidor Rabi, C. C. Lauritsen, and Hartley Rowe, technical advisor to General Eisenhower. Oppenheimer was free to place these men into sticky problem areas.

One special committee created by Oppenheimer was the Intermediate Scheduling Conference formed in August 1944. It was to coordinate the activities, plans, and schedules of groups that were formally or peripherally concerned with testing the implosion bomb. Parsons acted as chairman, and in November the committee was formalized with membership that included, among others, Bainbridge, Bacher, Kistiakowsky, and Norris Bradbury. Very shortly the committee became concerned with both the implosion and the gun bombs.

Another committee, called the Technical and Scheduling Conference, was created. In December 1944, it assumed responsibility for scheduling experiments, shop time, and the use of active material. Oppenheimer called Sam Allison back from Chicago to head the new advisory body. Allison was an experienced physicist and could bring a freshness to the staggering demands for coordination between many groups in the Laboratory. Each "conference" was called to address a particular problem area, such as the development of explosive lenses or experiments with the precious U 235. The conference had no fixed membership or topic. In many ways, it assumed the duties of the Technical Board.

The more intricate problems of scheduling the implosion development program were given to the Cowpuncher Committee. Euphemistically named to "ride herd" on implosion, the committee consisted of Allison, Bacher, Kistiakowsky, Lauritsen, Parsons, and

Rowe. The committee began meeting in March 1945, and followed the progress of experiments and developments on implosion. In April 1945, when the committee recommended freezing the design of the bomb, the members began monitoring and approving concurrent work within divisions. They became, for example, responsible for the tamper design work in G Division.

In March 1945, as both weapons moved toward completion, Oppenheimer created an interdivisional Weapons Committee to absorb the work of the Scheduling Conference. The new committee, with Ramsey as chairman, was responsible directly to Captain Parsons. It included, A. F. Birch, Brode, Bradbury, Lewis Fussell, Gene Fowler, and Sam Morrison. This committee was responsible for the delivery of combat weapons.

Also in March, Oppenheimer created two new organizations with the status of Divisions, the Trinity and Alberta projects. Trinity had become the code name for a proposed test of the new Fat Man weapon. Alberta, or Project A, was directed by Parsons and began planning for the overseas use of the bomb. At the very end of the war, Oppenheimer created Z Division to handle engineering and production needs concerned with adapting airplanes to use the weapons.

One of the major effects of the reorganization was to free personnel for reassignment within the Laboratory. Already there was a pressing need for more men in almost all division programs. Recruitment had fallen off because of the Laboratory's poor hiring program, and much of the recruitment was still by person-to-person contact or through reassignment of Army draftees. Oppenheimer also believed it was due in part to a lack of understanding among his project leaders as to what their specific needs were. Oppenheimer told his senior staff that personnel should be matched with the job by both research experience and temperament. There was a particular need for PH.D.'s, although as Oppenheimer pointed out, a B.A. in 1939 was equivalent to a Ph.D. in 1944. Oppenheimer was not unaware of the problems facing the Laboratory in the next year. The European war appeared to be ending, and the country as a whole would be likely to have a feeling of returning to normalcy. Good scientific men would want to be able to move back into industrial jobs or university settings. The problem was complicated because Los Alamos salaries were not competitive with those of industry.[3] Key Laboratory men such as Bacher and Shane were again asked to undertake lengthy recruitment trips to find new personnel in other parts of the Manhattan District. Eventually the effort netted men from Oak Ridge, the Metallurgical Laboratory, and laboratories in New York. Even a few more Army scientists were located and reassigned.

Oppenheimer also learned in the fall of 1944 that a rocket research project, called CAMEL, at the California Institute of Technology was nearing completion. The staff appeared to offer excellent possibilities for Los Alamos. Oppenheimer met with C. C. Lauritsen, the project's director, and asked about the possibility of collaborative work. After consultation with Groves, Conant, and Bush, Oppenheimer established an arm of the Los Alamos Laboratory at Cal Tech and kept the name CAMEL.

CAMEL studied problems associated with bomb assembly and delivery. Some work began immediately on implosion assembly, lens mold design, fuzing, and assembly of the high-explosive components. CAMEL also prepared Fat Man mockups called pumpkins, and dropped them from airplanes to test for stability. Lauritsen had a seat on the short-lived Cowpuncher Committee.

The continued expansion of the Laboratory still pressed every corner of Los Alamos life. The great tension and acceleration of work produced a momentary backlash in mid-1944. There was a drop in morale. Numerous complaints were still fed every day into Oppenheimer's office about wages, housing, working conditions, excessive hours, and the like. Groves had already ordered Colonel Ashbridge to cancel leaves for all military personnel at Los Alamos and to order a 54-hour work week. Groves had always believed the scientific staff to be on an "unlimited time" basis. Oppenheimer met with his new Administrative Board on August 3 and told them that the problem was not so much one of discontent among technical workers, but rather that they were not working very hard. It was an unusual attack from the normally patient director.

Oppenheimer cited the example of "certain" lackadaisical chemists being transferred from several divisions within the Laboratory to Kistiakowsky's implosion program. He then listened to another round of complaints, and agreed to identify several people within the laboratory to whom complaints could be brought. Then he announced that the Tech Area siren would sound at 7:25 A.M., 7:30, 8:25, 8:30, 12:00 noon, 12:55, 1:00, and 5:30.[4] This was to act as an inducement to be at work on time and to remind people of the lunch hour and closing. Much of the malaise in Los Alamos could be attributed to the quandary over implosion and the effects of rapid and unplanned expansion. By 1944, there were over 2,000 scientists and technicians at the Laboratory, most of them with families. Everything was hard pressed: housing, laundry, water, and most home amenities. The Laboratory was suffering too, with an antiquated wage and salary plan, poor procurement, waiting lists for critically needed staff, and poor management within groups of staff members. The August reorganization was like a shot in the arm.

Community problems remained, despite Oppenheimer's administrative juggling. Housing, for example, was one of Oppenheimer's most frustrating and seemingly unresolvable headaches. The tentative distinction between "scientist" and "technician" continued to put the better housing into the hands of the scientists. The McKee Construction Company finished its third phase of housing in December 1944. It was as if nothing had been built. The Laboratory finally tried offering technicians a $100 bonus if they came without their families. At least part of Oppenheimer's headache was caused by military perceptions of Los Alamos as a purely temporary effort. Housing was a major investment, and the Army was reluctant to spend more money than necessary. Groves continually dragged his feet on Los Alamos requests for more housing appropriations.

The fact remained: housing was critical. Over 200 men were recruited in November and December of 1944. One third resigned by January, stating hardships involved in living and finding suitable housing in Los Alamos. The Administrative Board stepped in with a suggestion that all housing be given out on an "availability basis" as the individual's contract was negotiated. A married man with two children would be entitled to a two-bedroom apartment, for example, but single men would have to accept less. The board dryly noted that there were "differences in the quality of housing."[5]

Oppenheimer was never able to resolve the housing problem satisfactorily. In one sense, he may have agreed with Groves and the military that Los Alamos was only a temporary place and only a Camelot in spirit. Oppenheimer believed that the real problems—and challenge—always lay within the mind.

11. THE SECOND YEAR, 1944-1945, PART II

Oppenheimer's reorganization coincided with new pressure from General Groves. Los Alamos, like most Manhattan District projects, was behind schedule. In April 1944, Groves suggested to Secretary of War Stimson that the 1942 goals of a bomb by the end of 1944 or early 1945 might still be met. Five months later, in August, Groves was forced to amend his predictions. The bomb, he wrote, depended on the production of uranium and plutonium and on "experiments yet to be conducted."[1] The "experiments" to be conducted referred to work in Los Alamos on implosion.

There were three weapons envisioned by Groves and Los Alamos: MARK I, as it was called, was a low-yield, low-power uranium gun bomb; MARK II was a more sophisticated uranium gun with greater explosive power; and MARK III was an implosion weapon that required relatively little fissionable material but gave the greatest explosive force. Groves wrote Stimson again and assured him that a MARK II weapon would be available with certainty by August 1945. If implosion experiments were successful at Los Alamos, a MARK III bomb would be available sometime between March and June of 1945.[2]

The implosion bombs described by Groves were seen as low-yield bombs, weapons with TNT equivalents of perhaps several thousand tons. Such weapons were capable of what Groves called "Class B" destruction: extensive damage that was beyond repair to perhaps 75 percent of all the buildings in an area of five or six square miles. The MARK II bomb—a uranium gun bomb—would be capable of twice as much explosive power and damage as a Class B bomb.

Groves made his report on the basis of Oppenheimer's projections for the Laboratory. Oppenheimer had found it necessary to freeze weapons design in February 1945. Groves was adamant about meeting a summer deadline for delivery of a gun and hopefully an implosion bomb. There was pressure on Groves, of course, from Stimson and ultimately from Roosevelt. The events in Europe suggested the war was close to being over. Groves desperately wanted to see the bomb's use in at least one theater of war.

The freeze left Edwin McMillan and his Gun Group with responsibility for the uranium gun program. Kistiakowsky's staff believed that a U 235 implosion bomb was not possible without considerably more research. Further calculations and experiments continued to dim the possibility throughout early 1945. Cross-section data, for example, indicated that U 235 would be considerably less powerful than plutonium in an implosion weapon. McMillan's work sought to complete the uranium gun in time for Project Alberta to prepare it for field use.

The February 1945 decision to freeze weapons design was the result of Oppenheimer's long consultation with his staff and with General Groves. In November 1944, Groves gave his go-ahead to freeze designs. For implosion, this meant a major thrust toward refining the explosive lenses. Scientists had considered several shapes for the explosives and several configurations for surrounding the plutonium and uranium tamper. Only two shapes emerged as clear possibilities in late 1944. Oppenheimer wanted full-scale lenses to be delivered by April 2, 1945, and tested along with the electronic detonation system by April 15. To meet an even more pressing deadline, Oppenheimer called for final fabrication of high-quality lenses for an experimental test of Fat Man by June 4. The detonation system itself was to be completed and tested by early June.

The Cowpuncher Committee had its hands full trying to bring together all the diverse lines of implosion work scattered throughout the Laboratory. Each scheduled date for delivery and testing met with delays caused by new problems. The metal lens moulds for making the explosive charges were not completed until May because of procurement lags in California. Delays in producing lenses for experimental testing meant delays in perfecting the timing and detonation systems. Final testing of the plutonium core, formed as two hemispheres, was eventually stalled for over two weeks by a sheer lack of personnel in the Metallurgy Division. There was a particular shortage of trained men among the teams engaged in explosives research. The result was a considerable cropping of the time that would normally be allotted for final engineering tests and modifications. The implosion bomb would have to evolve, much to everyone's uneasiness, from a stunted development cycle.

Hans Bethe's Theoretical Division remained somewhat outside the frantic machinations of implosion. More than any other, Bethe's division represented the academic scientist. Their offices were removed from the activities at Anchor Ranch and S Site. Safely ensconced near Oppenheimer's offices, they arrived like everyone else at 7:30 or 8:00 in the morning and conducted what appeared to be casual and unrelated conversations. Their blackboards were always left filled with calculations and notations at the end of the day. Everyone had to make a point of leaving "Do Not Erase" signs for the cleaning crews. There was always a measure of serenity in their cerebral work.

Bethe's staff had moved from concentrating on efficiency and critical mass to being involved in gaining a better understanding of the many forces inherent in an implosion bomb. There was still some concern with the former questions, but more and more they tried to grasp the hydrodynamics of implosion. The newly acquired IBM machines helped tremendously to cut down on time. Serber's Diffusion Group, for example, was able to assist McMillan in calculating the uranium gun's expected efficiency and probabilities of detonation. Peierls and his staff sought to develop a mathematical model for an "ideal" explosion.

There were other questions: What was the effect of extreme temperature rise during the implosion process? Could shock waves in fact be forced to converge accurately on the plutonium core? The problem was like having two hands uniformly crush an egg. Bethe's work was hampered by a lack of hard scientific data upon which to draw. As a result, it was necessary to rely on indirect evidence from experiments and on good guesswork. The real proof of their work would be with a successful test of the Fat Man.

As pressure mounted for such a test, the division was given additional tasks by Oppenheimer to relieve the load on other teams. A key component in the bomb was the initiator, the small neutron source at the very center of the plutonium core that would release neutrons at precisely the right moment. Work on the initiator seemed dangerously behind. Oppenheimer asked the T Division to pick up the task. Long mathematical programs were run on the IBM computers and produced a technical design by April 1945.

Oppenheimer also asked the Division to calculate the explosive force, or yield, of the Fat Man. Using data from different sources within the Laboratory, Bethe was able to predict a yield somewhere in the range of 5,000 to 13,000 tons of TNT.* The explosion itself was conceived of in terms of shock waves, a radiation "discharge," and an air blast that would be created by the explosion itself. Information on the explosive force of the weapon was necessary in order to design instruments for measuring different aspects of the explosion. If the prediction was too high, the instruments would fail to register properly, if at all; too low a prediction, and the instruments would be destroyed. The margin for error was flexible only to a limited degree.

In another part of the Laboratory, Robert Wilson's Research Division was similarly preoccupied with implosion. For a year some staff members had worked on the nuclear specifications of an implosion weapon, and on how to assemble the bomb's components and make them work reliably and all at the same moment. With Oppenheimer's major August reorganization, the division was reshuffled. Some staff went to Bacher's newly created Gadget Division, while others went to Fermi's Water Boiler Group. Wilson was promoted to head the new Research Division.

Much of the new division's work was to center on implosion. Emilio Segrè had just completed his study of spontaneous fission in plutonium and confirmed the presence of the 240 isotope. Wilson appointed John Manley and John Williams to conduct ingenious experiments to gather new data on neutron scattering and fission cross sections. Their results would provide the meat for further design of bomb components.

The August reorganization also brought Enrico Fermi as associate director. Fermi's arrival from Chicago reflected his value, as well as Oppenheimer's strong personal belief that he should be at Los Alamos.

---

*The actual yield of the first Fat Man was closer to 17,500 tons of TNT.

Groves had cautioned Oppenheimer, questioning the advisability of having Fermi continue as consultant to the Laboratory. He had even asked Oppenheimer to discontinue Fermi's visits altogether. Oppenheimer considered this attitude indefensible, and was required to raise the issue with his Governing Board on June 17, 1944. Seeking support from them, Oppenheimer told the members that he could not explain Groves' view nor could he support it. In fact, he wanted Fermi at Los Alamos. The board strongly concurred, and a plan evolved to bring Fermi to the Laboratory, escorted by Oppenheimer. It was left to Oppenheimer to mollify Groves.[3]

Groves' objections were never explained. Fermi arrived to assume general responsibility for all theoretical and physics research and was given F Division to head. The new division became an administrative catchall into which Oppenheimer could drop activities and problems that did not bear directly upon gun and implosion work.

Edward Teller was one such case. Teller was a man of considerable intellectual skill and personal complexity who did not work well with many members of the Laboratory's staff. He was nevertheless regarded as gifted and imaginative by most scientists and left mostly to himself, to pursue his speculations on the superbomb. Teller's ego was at least a match for Oppenheimer's, and their relationship grew increasingly strained as the work wore on. Oppenheimer was pressed on all sides for his time, and Teller had to settle for brief meetings once or twice a week—a considerable difference from their days together at Berkeley before the war. Teller's relationship with his own superior, Hans Bethe, also became strained, despite their long friendship. At one point, Bethe asked Teller to undertake a study of implosion hydrodynamics—a critical issue at the time—and Teller refused, citing his work on the Super.[4]

Oppenheimer was forced to put Teller in an independent position in June 1944. Oppenheimer valued Teller's intellectual skills, but was sensitive to his personality. With the arrival of Fermi—a man Teller could respect as his equal—Teller was placed in charge of a new group called The Super and General Theory, and was given some additional staff members. Teller had support outside the Laboratory, however. The prospect of such a powerful weapon could not fail to interest Groves and Tolman, and both men viewed the prospect of the Super as momentous. Groves noted that "'the Super cannot be completely forgotten if we take seriously our responsibilities for the permanent defense of the U.S.A.'"[5]

In the beginning, Teller could do no more than cope with the theoretical concepts for such a weapon. Not until 1944 was Teller's team able to speculate on more practical concerns, such as using liquid deuterium and ordinary fission bombs for detonation. Work became more intense in the spring of 1945, when Teller was able to announce that a Super might be capable of generating forces as high as ten million tons of TNT.

As a corollary, Teller was able to speculate on the Super's potential for destruction. Such a superbomb would not be the largest explosion witnessed on earth, however. The Laboratory already had combed historical records for large explosions, such as volcanic eruptions and meteorite impacts. The eruption of the Krakatoa volacano in 1883 was thought to have generated an explosion heard 3,000 miles away. The large meteorite craters in Arizona and Siberia gave evidence of explosions larger than a ten-million-ton Super.

On a relative scale, Teller was thinking that if an ordinary fission bomb could produce a blast of 10,000 tons of TNT in an area of 10 square miles, a Super could produce a similar effect over a thousand square miles. The effect of the Super would easily be sufficient to devastate New York City, for example. The next step was to theorize how such a blast might be increased or diffused over a larger area by altering the detonation site or the altitude.

It was speculated that more damage would occur if a Super could be placed underground or underwater near a continental shelf. Such a location would have the effect of a large earthquake. Even more exciting, to some at least, was the thought that if a Super could burn a 10-meter cube of liquid deuterium at the height of 300 miles, the effect would be one of 1,000 "ordinary" Supers detonated at the height of 10 miles. The potential damage staggered even Teller, for such a weapon would lay waste to a million square miles.

Less dramatic issues were addressed by Fermi's other teams. The cloudy prospects of implosion in late summer 1944, resurrected the possibility of perfecting the "autocatalytic" assembly bomb. This process had been given up early in the year when its efficiency seemed considerably lower than that of the gun and implosion bombs. The autocatalytic process employed small pieces of material to absorb neutrons at the right moment in the active core. One possibility was to coat small pieces of paraffin with boron, a high absorber of neutrons that was used in reactors, and to place them within the fissionable plutonium to keep the entire assembly subcritical. By forcing a chain reaction, the boron bubbles would be compressed and reduce the neutron-absorbing capacity and raise the level of criticality. Further experimentation, however, revealed the process to be lethargic. The explosion—if one occurred—would be minimal.

Fermi's work with the Water Boiler brought good news. With its operation many unresolved questions were answered. The reactor was called the Boiler, not because it actually boiled water, but because water was used to cool the system. Cold water was pumped into the reactor to cool the nuclear core, and was extracted as boiling water. In one experiment, a team of thermal neutrons from the Boiler was fed onto a small target of U 235 that was placed within a small sphere of uranium. The fission process that occurred within the sphere was measured, and the Laboratory was finally able to announce the critical mass necessary to sustain a supercritical reaction. Piece by piece, the Laboratory began to take empirical evidence in hand.

Most of the Laboratory's applied work and its resources were centered in three divisions: Ordnance, Weapons, and Explosives. As a result of the reorganization, Parsons was freed from implosion research *per se*, and could now give more attention to producing field weapons. Beginning in September 1944, Parsons pursued two major lines of work: completion of earlier bomb designs, and conclusion of testing programs to permit final delivery to the Army. In March 1945, a team called Project Alberta began to assume responsibility for delivery. While Parsons had general overall responsibility for both gun and uranium weapons, Kistiakowsky's X Division concentrated solely on perfecting the Fat Man implosion bomb.

The basic design of the uranium gun, called Little Boy, had been worked out in the first half of 1944, and the information on the critical mass from Fermi now permitted final design of the active components. The gun barrel was not unusual in terms of ballistics but was somewhat odd in appearance. It weighed only about half a ton and was now reduced in size to six feet. On its muzzle was a large thread. There were originally two types designed, called Types A and B. Type A was of high-strength alloy steel, was not radially expanded, and had three primers around the barrel. Type B was made of ordinary steel, radially expanded, but with the primers inserted in the nose of the breech. The Type B gun eventually was elected because it was lighter in weight and because the radial expansion was taken as an excellent test of the forging quality.

As a weapon, one end of the barrel contained explosive charges and approximately one half of the Uranium 235; the other end of the barrel was fixed and contained the remaining U 235. Although both pieces were machined and roughly shaped as cylinders, one half consisted of a male plug and the other a female receptacle. Both pieces were fastened by bolting or screwing to either end of the gun barrel. It was a surprisingly simple mechanism that belied its actual expense: the uranium in the gun represented an investment of almost a billion dollars.

The Fat Man was more complicated. Despite the lack of tested explosive lenses, the Ordnance Division had to proceed with a rough design of the bomb in order to plan for delivery. While the actual size of the plutonium core was about that of a grapefruit, the uranium tamper and explosive charges added considerably to the bomb's final size. The core, tamper, and high explosives were held in place by a metal sphere made of twelve pentagonal sections. These were bolted together to form a sphere that was in turn protected by an egg-shaped outer shell. Stabilizing fins were attached and, altogether, over 1,500 bolts were used to assemble the final weapon.

By the end of the war, a newer Fat Man was developed that was easier to assemble. It consisted of a spherical shell made from two polar caps and five equatorial zone sheets. This, in turn, was surrounded by an ellipsoidal shell of armor with stabilizing fins. All of the electrical detonating and fuzing equipment was mounted in the space between the bomb sphere and the outer shell.

Robert Brode and his men were still stalled with problems in the Archie fuzing mechanism. The units were very complex and were failing to meet reliability standards. Archie was a tail-warning device that employed a radar system to close an electrical circuit at a predetermined altitude above the target. Four such Archies were used with each fuze—in effect, this was a network of relays, arranged so that when any two of the units fired, the mechanism would send a firing signal to the next stage. The second stage consisted of a bank of clock-operated switches that were started by arming wires pulled out of the clocks when the bomb dropped from the airplane's bomb bay. These clock switches were not closed until 15 seconds after the bomb dropped from the plane. Their purpose was to prevent detonation in case the Archie units were fired by signals reflected from the airplane.

A second arming device was a pressure switch that did not close until the bomb reached an altitude of 7,000 feet. In the Little Boy weapon, the firing signal went directly to the gun's primers; in the Fat Man, the signal activated an electronic switch which in turn closed the high-voltage firing circuits. A standby fuzing system also was developed. This unit, called Amos, was another radar device developed by the University of Michigan. The Amos unit was developed primarily for a bombing strategy requiring a lower altitude.

The height from which to drop the bomb was a question taken up by Shapiro's Water Delivery and Exterior Ballistics Group.

This team worked with John von Neumann to conduct a series of small experiments to investigate the effects of explosions in shallow water. Small charges were ignited in water depths from a few inches to a few feet. The results were projected for comparisons with fission weapons. These results established, contrary to expectation, that a surface explosion produced larger gravity waves than a sub-surface explosion of the same size. This evidence suggested that the bomb could be used effectively against harbors or ports.

Robert Bacher's Gadget Division continued its implosion work, and in April 1945, assumed responsibility for producing the final tamper and active core for the Fat Man. Additionally, they worked with the Theoretical Division to design and perfect the polonium initiator, as well as the electrically-activated detonators. Like the chemistry and metallurgy teams, G Division was soon overcrowded, and Oppenheimer authorized a new building in the Technical Area called Gamma. Already the division had been forced into the canyons to look for new testing and firing sites. In addition to S Site, where the explosive lenses were poured, the division built sites called Omega, Alpha, Beta, P, and X.

One of the most intriguing and dangerous groups in the Laboratory was Otto Frisch's Critical Assemblies team. At Omega Site, where they shared the facilities with Fermi's Water Boiler, the team prepared and tested the U 235 and Pu 239 cores which they had nicknamed the pit assembly. Frisch's men performed some of the most dangerous work during the war. Very little was known about critical masses or even about handling large quantities of uranium and plutonium. Experimentation with these materials was embryonic and required both scientific zeal and a large measure of daring and foolhardiness.

One of the most bizarre experiments was called tickling the dragon's tail, or simply, the dragon. In this experiment, a small slug of uranium hydride was dropped through the open center of an almost critical collection of similar slugs. The single slug was guided through by four rails and for a hundredth of a second the entire assembly would become supercritical and release heavy doses of neutrons. The reaction would quickly die down as the slug fell through the lowest layer. In that brief moment, however, the small pile would generate nearly twenty million watts of energy and almost instantly raise the temperature to fantastic levels. In February 1945, one such dragon experiment became so radioactively hot that the uranium slugs began to blister and melt. Work had to be stopped for several days.[6] One man, particularly, seemed fascinated by the experiment and conducted it several dozen times. His name was Louis Slotin.

Just a few months later, in June, a number of small uranium cubes were arranged within a six-inch polyethylene box. The box was being lowered slowly into a large tank of water when suddenly the material became critical. No one had expected such a reaction, and the tank had been designed with its drain valves some 15 feet apart. Before anyone could lower the water level, a chain reaction began and quickly intensified. No one was killed, but again the room became so hot that all experiments were canceled for several days.

The first death occurred in August 1945. Harry Daghlian, a young scientist working under Frisch, was assembling small bricks of uranium as a reflector around two plutonium hemispheres he had nicknamed Rufus. The purpose of the experiment was to build a wall carefully around the plutonium to a level just before it went critical. Each brick weighed just over six kilograms. As Daghlian began to lay the last brick, he noticed from nearby neutron counters that the entire assembly was on the verge of becoming supercritical. As he moved his hand away, the brick slipped and fell into the center of the pile. At that moment, the additional brick provided enough neutron reflection to make the assembly supercritical. A blue glow immediately burst from the pile as Daghlian pushed the brick off.

The immense radiation Daghlian received in that brief moment was deadly. Very shortly, second-degree burns developed on his body where the radiation had been most intense on his hands and abdomen. A fever developed, and after two weeks the burns blistered and he lost his hair. He died 28 days later.[7] The world had its first nuclear fatality.

Within a year, Louis Slotin would become the second victim of an accident while he worked with the same plutonium hemispheres. In Pajarito Canyon Laboratory, the plutonium was surrounded by a heavy beryllium shell some nine inches in diameter. The lower half of the beryllium tamper was already in place, and Slotin was slowly lowering the top half to the point where it was prevented from touching the bottom portion by the tip of his screwdriver. Somehow, Slotin let the screwdriver slip and the two tamper halves fell together. There was the same piercing blue flash and sensation of heat felt by Daghlian. This time, however, there were seven other men in the room. Fortunately for the others, Slotin's body absorbed most of the radiation and only he died. His body had been exposed to over 900 units of gamma radiation, and his death took a mere nine days.[8]

Bacher had appointed Luis Alvarez to lead the development of the electrical detonators. Even as early as 1943 it had been known that all the detonators had to fire simultaneously; it was particularly critical that all of the implosion bomb's explosive lenses be ignited at the same moment. Alvarez soon discovered that there was no commercially produced detonators with the degree of quality and reliability needed. Moreover, even if such detonators were available, an electrical system for firing each one at the same precise moment was not available. Alvarez also needed to develop a high-voltage power supply and a switching mechanism that would produce the right degree of simultaneity. The work was straightforward but laden with technical bugs. Many of the newly-developed detonators would fail in tests, and simultaneous firings eluded the scientists. In two years, Alvarez and his staff had prepared thousands of detonators of different styles, but not until May 1945, did they find the right materials for dependable units.

Responsibility for successful implosion results fell inevitably to Kistiakowsky and his X Division for incorporation into bomb design. Upon his appointment as division director, Kistiakowsky appointed Major W. A. Stevens as his assistant for administration and construction. Among his scientific staff, he appointed Norris Bradbury to be in charge of implosion research; Kenneth Bainbridge in charge of development, engineering, and testing for implosion; and Army Captain Jim Ackerman in charge of the development and production of explosives used in the Fat Man weapons. Within a few months, Kistiakowsky added new groups for overseeing the production of lens molds and electric detonators.

Oppenheimer was clear in his directive to X Division: conclude investigations into the high-explosive lenses for the implosion bomb, prepare for production of success models, and produce sufficient quantities. In actuality, the purity of administrative lines soon began to obscure. Kistiakowsky and Bacher had overlapping responsibilities for the implosion bomb. Bacher was concluding experimental studies, while Kistiakowsky was responsible for the development of bomb components. As they completed work, however, they sent their products to Parsons' Ordnance men for fabrication into manageable field weapons.

Kistiakowsky had his hands full with perfecting the explosives for Fat Man. In 18 months the division had made over 20,000 explosive casings and at its peak, S Site alone consumed over 100,000 pounds of explosives in one month. One of the division's major accomplishments was the development of methods for casting high explosives into the appropriate shape. Commercially available explosives

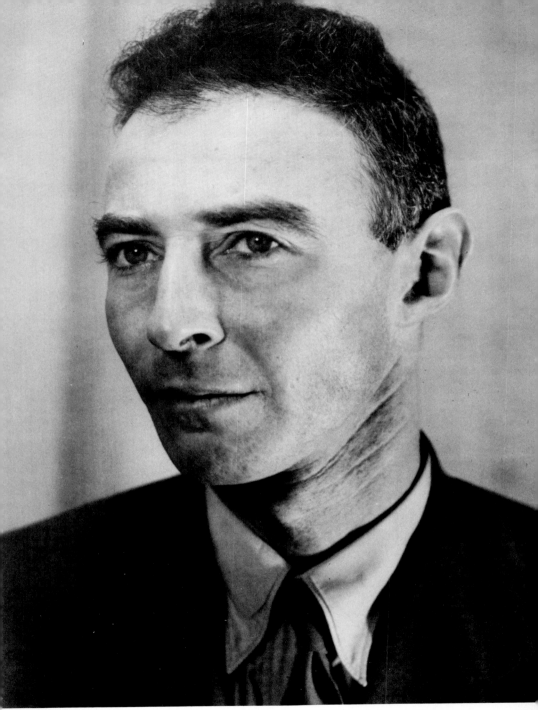

J. Robert Oppenheimer—Oppenheimer was director of the Los Alamos Laboratory from its creation in late 1942 until October 1945. This photograph was taken during the war years.

Navy Captain William S. Parsons—Parsons was an early member of the Los Alamos staff. He directed the Ordnance Division. He had a major role in the development of the "Little Boy" uranium bomb.

Major General Leslie R. Groves—General Groves was the overall director of the Manhattan Engineering District. He made Los Alamos the vital center of the Manhattan Project for the design and construction of atomic bombs. This photograph was taken shortly after Groves received his promotion from brigadier to major general.

George B. Kistiakowsky—Kistiakowsky was Russian-born and joined the Laboratory in the summer of 1944. He was made responsible for completing the faltering "Fat Man" implosion bomb project.

Kenneth T. Bainbridge—Oppenheimer gave Bainbridge the overall responsibility for planning and conducting an experimental test of the first atomic bomb. He labored for over a year preparing for the final moment at Trinity, July 16, 1945.

Edward Teller—Teller was also an early member of the Laboratory staff. He devoted much of his attention to the theoretical conditions needed for a fusion, or thermonuclear, bomb. He was later given the title of "Father of the Hydrogen Bomb."

Meeting of the Laboratory's famous Colloquium—These meetings were heavily protested by General Groves who wanted Los Alamos scientists to be compartmentalized for security reasons. First row, left to right: Norris Bradbury, John Manley, Enrico Fermi, and John Kellog. Second row, left to right: Robert Oppenheimer, Richard Feynman, and Phillip Porter.

Oppenheimer's resignation and awards ceremony—General Leslie Groves (center) presents Oppenheimer (left) with a Certificate of Appreciation from the Secretary of War. To the right of Groves is Gordon Sproul, President of the University of California. October 16, 1945.

Graduating seniors of the Los Alamos Ranch School for Boys—Graduation was held on the front porch of the school's main building, Fuller Lodge. This building was later used by the Laboratory as a hotel. Photograph taken during the 1920's.

Common Room at the Los Alamos Boys' School—Older boys had a common room for use after classes or at night. The school's official uniform was modeled after that of the Boy Scouts.

Early road to Los Alamos—This was a portion of the road to Los Alamos that greeted new arrivals. General Groves was finally forced to widen and pave the road in 1943. The original road to the Boys' School was even rougher.

The Los Alamos "Technical Area"—The Tech Area was the central complex of laboratories and offices for Los Alamos. The dirt road was the main artery into Los Alamos. On the right is Gamma Building, used for metal fabrication; to the upper left is T Building, used for Oppenheimer's administrative offices.

Administration Building—T Building was the first completed in the Tech Area. It housed Oppenheimer's offices, the Theoretical Division, a library, and a classified documents vault. The walkway connected with Gamma and E buildings.

Covered walkways over main road—These two walkways spanned the town's principal road. P Building is on the left and one of the Area Pass Offices is on the right. The road remained unpaved and unnamed until after the war when it became known as Trinity Drive.

Cyclotron Building—Building X was constructed early in 1943 to house the cyclotron. Construction personnel did not know what a cyclotron was, but were told to plan for a "large machine." The view is south into Los Alamos Canyon.

Arts & Crafts Building—The new Laboratory appropriated all the buildings of the Los Alamos Boys' School including the Arts & Crafts Building. The Jemez Mountain Range is barely visible over the roof of the building.

Los Alamos apartment buildings—Typical of early construction, these buildings contained three-room apartments. The small white plaque on the left end of the building gave a number for identification purposes. These numbers served in place of street names.

Luxurious apartments—These apartments built by the Sundt Company were considered the best in Los Alamos. Important scientists like Fermi, Teller, Bethe, and Allison lived in these two- and three-bedroom apartments with their families.

Life in Los Alamos—With the great influx of families, the Laboratory turned to the quonset hut known as a "Pacific Hutement." The Army built 56 of these before the end of the war.

McKeeville—This area, like others in Los Alamos, took on the name of its builder, McKee Company. Over 100 flat-roofed McKees were built.

Handling nuclear materials—Techniques for handling large quantities of uranium and plutonium were unknown in the early 1940's. Los Alamos scientists stored some of these materials in temporary outside buildings.

Jumbo on special transport—Built to contain the Fat Man bomb, the 200-ton steel vessel was brought by special train to a railroad siding at Pope, New Mexico. From there, Jumbo was carried across the desert by trailer. At the last moment it was discarded and placed several hundred yards from Ground Zero.

Los Alamos from the air—By the summer of 1945, the Laboratory had spread over most of Los Alamos Mesa. The pool in the center of the photograph is Ashley Pond. The Administrative T Building is in approximately the center of the photo.

Base Camp at Trinity Site—Base Camp was a small town built at Trinity to house the several hundred scientists who prepared for the first test of an atomic bomb. The buildings were largely for housing, makeshift laboratories, and storage. In 1977, only two stone buildings remained.

Jumbo in special tower—With increased confidence in the Fat Man bomb, scientists decided not to use Jumbo directly. A special tower was built several hundred yards away to hold the vessel. Scientists wanted to see what would happen to it during the explosion.

Fat Man tower at Ground Zero—Specially built to hold Fat Man and its detonation system, the tower rose 100 feet above the New Mexico desert. The large crate at the base contains the implosion bomb without its plutonium core. Electrical cables run from the small building at the top to electronic test equipment.

Fat Man at base of Trinity tower—Men assemble the "Gadget" at the base of the tower. The bomb has been preassembled with its explosive lenses and tamper in place. Only the plutonium core and initiator need to be inserted. Later the Fat Man will be hoisted to a small cabin at the top of the tower.

Interior components of Fat Man—Mock-up of implosion Fat Man bomb. Spherical shell is partially removed to reveal two rows of explosive "lenses" and the uranium tamper. Inside of the tamper are the plutonium core and polonium initiator. A cork liner protects the explosives from the metal shell. Systematically placed around the lenses were 32 detonators (not shown). The final Fat Man used at Trinity and in Japan differed slightly. This photograph has been highly classified for 30 years, and is printed here for the first time.

Final Fat Man dropped on Japan—Use of this bomb against Japan necessitated a B-29 bomber with enlarged bomb bays. The complicated electrical and detonation systems were packed into the front and rear pockets of the ellipsoid shell. A small removable panel on the opposite side permitted the bomb to be "armed" while in flight. With fins, the bomb was twelve feet long and sixty-two inches in diameter; it weighed 10,000 pounds. Fat Man was used against Nagasaki on August 8, 1945.

"Little Boy" bomb—This weapon used the "gun" method, in which one piece of uranium was fired into a second piece at the end of a gun barrel. Little Boy was considerably more simple in design than the Fat Man implosion weapon. It was ten feet long and twenty-eight inches in diameter; it weighed almost 9,000 pounds. Little Boy was used against Hiroshima on August 6, 1945.

were used, such as Composition B, Torpex, Baronal, and Baratol. The melting, shaping, and machining of these materials was particularly dangerous work, and new methods had to be found for handling the materials while maintaining a very high degree of quality and purity.

The question of explosives raised the possibility of a misfire. The quantity of explosives used in the Fat Man would be enough to scatter the precious plutonium in case the active material could not be made supercritical. This concern led to the development of several methods of recovering the plutonium in case the bomb failed to work properly.

Several ideas were proposed. One suggestion was to construct a water baffle that would contain the plutonium in case of a fizzle. Another possibility was to explode the device in the sand, and still a third possibility was to explode the bomb in a large solid container. Only the last idea seemed worth pursuing, and Robert Carlson designed a huge metal vessel that could withstand conventional explosive forces but would vaporize with a nuclear explosion.

The huge jug, called Jumbo, was an enormous steel cylinder 25 feet long and 12 feet in diameter. The Laboratory placed an order for Jumbo in August 1944, and received it at its test location in Southern New Mexico in May 1945. The entire device weighed over 215 tons and had to be transported across the desert in a special 64-wheel trailer. At the last moment, however, confidence in the Fat Man had strengthened. Plans to use Jumbo were dropped, and the vessel was left upright within a steel frame several hundred yards from the point of detonation.

Contingency plans for recovery, however, were continued. If perhaps Jumbo or some other device was used, there was still the chance of a misfire. Oppenheimer thought the Clinton Plant in Tennessee would be better prepared for plutonium recovery. If some nuclear reaction did occur and Jumbo was destroyed, the resulting material would be radioactive. Oppenheimer finally asked the Chemistry and Metallurgy Division to make arrangements at Bayo Canyon to handle decontamination emergencies. Several heated buildings, well spaced from one another, were constructed, along with an extraction column for recovery purposes.[9]

Joseph Kennedy's Chemistry and Metallurgy Division meanwhile had begun to receive quantities of uranium and plutonium and had made some interesting discoveries about the substances. Plutonium, for example, was discovered to have different allotropic forms (the same element in several different conditions) whose existence depended on the state of the metal during purification. This discovery explained difficulties experienced in working with plutonium, but by mid-1944 everyone felt comfortable handling the materials.

Los Alamos received uranium from the Y-12 plant in Tennessee as a purified fluoride (uranium hexafluoride). General Groves had forbidden transportation of the precious substance by air as being too risky. At 10:30 A.M. on predetermined days, Army couriers in civilian clothing would leave Knoxville with the uranium packed in ordinary luggage. At 12:50 P.M., the men would catch the Southland Express for Chicago, where they were met by members of the Manhattan District Office. At noon the next day, another team of Army couriers boarded the Santa Fe Chief for Lamy, New Mexico, the closest stop to Los Alamos.

In Los Alamos the fluoride was reduced to a metal form. To purify the metal, the Laboratory teams put the substance through a long and tedious extraction process. Sam Marshall's team would take the uranium and fabricate the necessary parts for the Little Boy weapon. A block away, Charles Garner's Plutonium Group would similarly fabricate the purified plutonium into matching hemispheres coated with a thin layer of silver. As the first test of the Fat Man neared, a series of small blisters and pinholes was discovered on the surface of the core. Since the test was only days away, there was no time for recasting, and Garner's men were forced to polish the sphere as best they could.

The Polonium Group had particularly dangerous work. Polonium was used in the initiator and possessed a dangerously high alpha radiation activity. It was virtually impossible to come into contact with the substance and not absorb it into the body. Like gamma radiation, it was deadly in heavy doses. Fortunately, polonium, unlike plutonium, is eliminated rapidly from the human system. The unusual element also had extreme mobility, and in many shipments of the element from a Monsanto plant in Ohio, the polonium would travel from thin platinum foils to the metal sides of sealed shipping containers.

The quantities of uranium and plutonium were precious. For a long time, what little nuclear material Los Alamos possessed was stored in the old stone icehouse built by the Boys' School. Kennedy himself raised the question of responsibility for the loss of U 235 or Pu 239. Oppenheimer asked that it become the responsibility of every person who released the material to another worker to be convinced that the recipient was capable of assuming responsibility. Kennedy also raised the question of providing guards at storage centers. Kennedy was less concerned with theft than with disasters such as fire. Although the Post military offered to guard the material, no provisions were made at the time for fire protection.

In January 1945, a fire did destroy the C-Shop building in the Tech Area. Everyone realized that a similar fire in the present plutonium facilities could mean a catastrophe for Los Alamos. A new building was ordered for the eastern part of the mesa. DP Site was to be a prototype of future Laboratory buildings. All construction was of noncombustible materials—steel walls and roof, rock wool insulation, and plaster lining. Even the ventilation system was revolutionary. It not only cleaned air from the outside, but processed it again before releasing it.

In Washington, Groves took note of the fire with alarm. He could ill afford *any* setback to the Manhattan effort. Los Alamos was only a part of his operations, but it was the kingpin of bomb development. At the moment he was equally concerned with upgrading the uranium and plutonium outputs at Oak Ridge and Hanford. In Tennessee he pushed his deputy, Colonel Nichols, to expedite the U 235 separation process. The low-grade production plants called S-50 and K-25 were urged to feed the sophisticated electromagnetic plant at Y-12 swiftly. Nichols was forced to exhort the workers in the name of the war effort. At Hanford, early delays seemed resolved with the addition of more uranium feeds into the reactor. By May 1945, production was up five times. Groves ordered his Hanford director, Colonel Franklin Matthias, to increase shipments to Los Alamos with all speed.[10] Unless a major disaster occurred, Groves felt that the deadlines reported to Stimson and the President could be met.

There was still trouble in other quarters, however. Congress was beginning to question the vast expenditures of the Manhattan Project. One congressman, Albert Engel, learned of another major request from the War Department for the Manhattan District. Enraged, he wrote Secretary Stimson a long letter of protest. Stimson was able to appeal to Engel's sense of loyalty and to quiet congressional fires for the moment. War Mobilization Chief James F. Byrnes wrote President Roosevelt a memorandum suggesting an impartial review of the overall project by scientists not associated with the Manhattan District. This review, he felt, might help settle congressional concerns that the money was being well spent. He noted that "the expenditures for the Manhattan Project are approaching two billion dollars with no definite assurances yet of production." While the impartial review might offend some scientists such as Bush, the "two billion dollars is enough money to risk such hurt."[11]

Roosevelt sent a copy of the memo to Stimson a few days later and asked for a meeting at Stimson's convenience. Stimson joined the President for lunch and was able to defeat the suggestion for a review by pointing out that the Manhattan Project included every major physicist of standing and four Nobel Prize winners. The only real solution was to have the bomb ready and a success.

As summer of 1945 neared, Oppenheimer could take stock of the fantastic growth of Los Alamos. The Laboratory had grown considerably from his original estimate of one hundred scientists and their families and a few buildings. The Technical Area alone contained 37 buildings and more than 200 more were scattered in nearby canyons and mesas. There were 620 rooms in more than 300 apartment buildings, 200 trailers, and 52 dormitories. And finally, the one hundred scientists had grown to more than 4,000 civilians and 2,000 military.

Oppenheimer could wonder when it would all produce the product with a bang.

12. THE HARVEST

A special excitement began to permeate the Laboratory during the early months of 1945. The events and activities of two years were gathering toward conclusion at Los Alamos. At last there seemed to be an end in sight. Spirits were never higher, the work never more intense than in the first half of 1945. The hardships of mesa living continued, but were softened by high expectations for the next few months and by the suspicion that mortal men were about to do something godlike.

The hard work was bearing fruit. The organizational shakeup of the previous August had finally moved Laboratory resources successfully behind the implosion approach. Kistiakowsky's leadership had brought new energy to implosion developments. Bethe was close to concluding his critical studies on plutonium and uranium, and Parsons had been freed from research demands of the plutonium bomb to concentrate on perfecting the uranium gun. He also needed to prepare the implosion and gun weapons for use by the Army against Japan.

In January, Fermi was able to announce some encouraging results on implosion studies. A small sphere of plutonium from the

Clinton reactor had been used in a series of experiments to determine the critical mass. The sphere itself was less than one inch in diameter, but it was useful at last in confirming theoretical speculations about the amount of plutonium needed for a bomb; five kilograms would come close.

Bacher's Gadget Group had also begun to make clear progress toward the perplexing problem of implosion: symmetrical shock waves. The basic principle behind implosion was the inward force created by exploding TNT charges surrounding the plutonium core and uranium tamper. The entire weapon was a sphere within a sphere within a sphere. At the very heart was the polonium initiator, the source of neutrons that would begin the chain reaction at precisely the right moment. The plutonium and initiator were carefully machined to fit exactly within the tamper, and several layers of explosive charges, called lenses, completed the outer sphere.

Bacher's problem was twofold: how to configure or shape the lenses and how—and from what—to manufacture them. Bacher had four of his group leaders working on the problem: McMillan, Neddermeyer, Rossi, and Froman. They had inherited the implosion theory developed by Neddermeyer in 1943, and later refined by von Neumann and Peierls during 1944. All four leaders and their groups worked very closely with Willie Higinbothan of the electronics team to develop a bewildering array of experiments to test and record the effects of different explosive shapes.

No one had ever studied explosions in a way that was useful to Bacher's men. As a result, entirely new methods of experimentation were devised. One method developed by Bruno Rossi utilized X rays to follow the effects of imploding explosions. Later flash X-ray photography was added to the repertoire. The days in Los Alamos were interrupted again and again with the sound of experimental explosive lenses being fired at Anchor Site some three miles away from the Tech Area.

Another experimental technique used a substance called radiolanthanum to trace the effects of imploding waves, and still another method recorded the electrical contacts between imploding spheres and prearranged wires. In mid-December, a series of experiments utilizing Rossi's techniques finally gave evidence of controlled symmetrical waves achieved through the use of carefully determined shapes. At K Site, a betatron was constructed to conduct experiments with cloud chambers and a maze of electronic circuitry. The betatron was a doughnut-shaped accelerator in which electrons were speeded up by a charging magnetic field. This machine could achieve many millions of volts of energy. In April, the complicated process gave the Laboratory a solid picture of the effects of imploding shock waves.

The new year also brought tentative news from Alvarez. Dependable detonators at last seemed to be in hand. Not only did a new model provide the needed electrical charge, but it appeared to fire dependably time and again. Experiments showed that simultaneity occurred evenly with dozens of detonators in less than a microsecond. Alvarez provided Rossi's group with the new detonators in order to put through an additional round of tests. Rossi's tests in January showed a vast improvement in quality and dependability.

A year's work on implosion was reviewed on February 17, 1945. Allison's Technical and Scheduling Conference held a Saturday meeting to review implosion work throughout the Laboratory. From S Site came the assessment that present capacity for producing lenses was severely strained. The molds and techniques used for casting these explosives were already strained to their limits. It was impossible to continue trying to develop multiple lens designs. Charles Lauritsen, who was Project CAMEL's director, attended the meeting and suggested that Los Alamos would have to decide on one line of lens development if the Laboratory were going to meet the deadlines set by Groves and Conant.

The evidence from Bacher seemed to suggest that two rows of explosive lenses around the core and tamper would offer the best promise of a uniform explosion. Each charge resembled a small pyramid with a flat top. Allison favored this single approach, while Oppenheimer favored pursuing both this approach and another alternative lens design. The problem, Allison argued, was that Los Alamos did not have the resources to continue developing both models.

Oppenheimer deferred a decision until he could consult with General Groves. A week later, Groves arrived in Los Alamos for a special conference with Oppenheimer and his division leaders. A compromise was worked out: Los Alamos would continue with the design favored by Allison, while the team at CAMEL in California would continue work on the second model. This seemed to be insurance against failure of one method or the other.

The choice of a single implosion design eased the strain on the Laboratory. Parsons, for one, was relieved. He could now complete work on both the implosion and gun bombs. Under his leadership, Project Alberta began to draw resources together for the final production of weapons for the military. Norris Bradbury and Roger Warner took charge of the Fat Man weapon, and Francis Birch took over the Little Boy. Robert Brode assumed responsibility for fuzing, Lewis Fussell took over the electronic detonator system for both bombs, and Philip

Morrison and Marshall Holloway assumed responsibility for the active material and tamper for Fat Man.

They would ultimately produce weapons compatible for use in Army airplanes. This involved developing special casings for the bombs, testing their aerodynamics in flight, and determining the best methods of release and activation from airplanes. At the same time, the Alberta teams began to plan for shipping the bombs and their support equipment overseas. The recommendation of the Review Committee in 1943 to pursue both research and bomb design at the same time now began to pay off. The exterior shells of both bombs had been prepared and tested at Wendover Field in Utah by late spring. With the Fat Man and Little Boy components nearly completed, it would be a relatively simple matter to assemble them along with their external shells.

In the Technical Area, the Chemistry and Metallurgy Division began a frantic around-the-clock race to fabricate the nuclear cores for both weapons. Theirs was the task of taking the uranium from Oak Ridge in its fluoride state and the plutonium nitrate solution from Hanford and reducing both to pure metals. Uranium was the easier to handle; plutonium the more difficult. In addition to refinement and casting, the division oversaw the construction of its new plutonium purification plant at DP Site.

The vast Manhattan District now turned its attention to delivering its creations to Los Alamos. The massive plants at Oak Ridge began at last to produce significant amounts of U 235. Hanford looked better: plutonium shipments in April were considerably larger than those of U 235, and shipments in May were five times as large. There would now be enough plutonium in July for a test bomb and a field weapon in August.

The hectic work in Los Alamos was stopped suddenly on April 12, when news of Franklin Roosevelt's death came over the radio. Oppenheimer sent word to all laboratories and invited everyone who could come to the flagpole in front of the Administration Building for a memorial service. He spoke for only a few minutes and drew from his reading of the *Bhagavad-Gita*. Oppenheimer spoke quietly with a voice that was hard to hear, but the effect was powerful and memorable. Many felt the strange spirit of the mesa for the first time.

April also marked the two-year anniversary of Los Alamos. It was the first good month for the Laboratory since 1943. Oppenheimer could see his diverse activities coming together. Frisch and Slotin were close to achieving the first critical assembly of U 235 in mid-April. The critical mass that for so long had been only theory could now be confirmed by experiment. Robert Wilson's men were also very close to determining the quantity of uranium to be used in the gun's target and projectile.

Oppenheimer reported the development to Groves. The uranium gun would be ready by August 1. Further experiments by Bacher brought even better news of symmetrical impolsions with the blunted lenses under development. Work on the initiator had resulted in three designs that looked promising, and Bethe expected to make his first choice by the end of April. Bethe's theorizations also suggested that the Fat Man would yield about 5,000 tons of TNT. This amount was better than previously thought. Only the acquisition of successful lenses and detonators seemed to darken the chances for meeting the July deadline of completing one Fat Man.

At the last moment the detonators proved unable to withstand the heavy testing. Despite the patient, thorough work by Alvarez, the first combat detonator systems experienced a new series of failures. The high hopes of January began painfully to fade with May. Edward Lofgren and his staff had built combat versions of Alvarez's experimental model. At first they seemed to work with unusual reliability; then inexplicably they developed failures. Lofgren threw his men back into redevelopment.

In other parts of the country, the contractor developing the firing circuits fell behind schedule in delivery. Testing the entire detonator package subsequently fell behind. Another contractor producing lens molds fell behind. S Site had recently completed new buildings to handle the increased production needs and found itself suddenly without key equipment. Everyone at Los Alamos found the July and August deadlines pressing with great speed.

June seemed to revive an atmosphere of promise. New molds finally arrived and were put to use immediately. Frisch's Critical Assembly Group concluded tests with the U 235 gun that suggested the gun would be safe from predetonation. This news allowed the Technical and Scheduling Committee to assign deadlines for the casting of the target and projectile. Lofgren produced a newly modified detonator with minimal levels of failure. Even the manufacture of initiators was accomplished successfully, despite a mishap when a technician accidentally dropped the first initiator into an open sewer pipe.

On June 24, Allison's Cowpuncher Committee met to review the critical mass experiments being conducted by Slotin at Omega Site. Tickling the dragon finally paid off. Slotin and Frisch could now tell the Laboratory the critical amount of plutonium needed for Fat Man. The Cowpunchers turned the information over to Eric Jette to cast and shape the small sphere.

Parsons also moved to complete both Fat Man and Little Boy for combat use. The Little Boy was well underway. The U 235 would be ready for use by July 3. The protective shell had already undergone flight tests in Utah, and appeared stable when dropped from the new B-29. Only the Fat Man lived up to its name: in flight the bomb was unwieldy. Parsons already knew of several modifications that would make it easier to assemble and improve its characteristics in flight, but unfortunately there was not enough time to undertake each change. Parsons would have to make do with what changes could be made quickly before the bomb was shipped to a small island base in the Pacific.

The Army had loaned part of its air base at Tinian Island to Los Alamos. From Laboratory plans, base construction had begun in April, and Los Alamos began to assemble equipment, special tools, and replacement bomb parts for shipment. On June 18, the first contingent of Los Alamos personnel left for the small island near Guam. Japan was just a step away.

Success on such a broad front gave Oppenheimer time to assess Los Alamos and to make tentative plans for the future. In early May he was able to summarize for Groves the state of affairs and the work yet to be done.

All the effort of Site Y, he said, was devoted to completing by summer the first models of the Fat Man and Little Boy. While there still remained work to be done, it was appropriate to think of the period of work in late summer and fall. Even with the end of hostilities, Los Alamos would need to continue the production of weapons so far developed. Unless there were plentiful supplies of U 235, it was clear that the implosion model would represent the overwhelming majority of bombs to be produced by Los Alamos immediately after the end of the war.

This assessment represented the Laboratory's view that the implosion process was superior in effectiveness to the gun process. Both uranium and plutonium could be used in the Fat Man design. Los Alamos would nevertheless maintain a store of Little Boy components. This seemed a wise protection against the ultimate failure of Fat Man or the possibility that the Army would find the Little Boy to be tactically superior in field use.

Oppenheimer also recommended that Los Alamos continue, for a while at least, the production of all necessary explosives, tampers, and initiators needed for both weapons. Oppenheimer hoped, however, that private industry could assume the responsibility for producing detonators and firing units. If this were possible, it would mean savings to Los Alamos of $25 million in 1945.

Oppenheimer sensed the need to prepare for the end of the war and the return of Los Alamos to more conventional life. He strongly urged Groves:

... that an effort should be made as soon as it is politically feasible to renegotiate contracts by which we are now obtaining components from outside vendors, and to provide a sound industrial basis for the fabrication, along the lines developed here, of explosive, active, and initiator components, so that the whole problem of producing gadgets, as opposed to the problem of their development, can be regarded as solved along standard industrial lines, and with no undue and improper dependence upon the existence of Site Y[1]

Oppenheimer was making a strong case for relieving Los Alamos of the production chores. In a more subtle way, he was suggesting that the role of Los Alamos was close to ending.

Much improvement of the current weapons remained, of course. The design and engineering of the Fat Man, for example, were clumsy and posed problems durings its release from the airplane. The most important refinement needed to be made in the tamper and active material core design. Los Alamos now believed that the Fat Man could be made perhaps three times as powerful by removing some two thirds of the uranium tamper. This reduction could then mean a corresponding decrease in the amount of plutonium needed. If both the core and tamper were reduced in size, the bomb would be less powerful, but more bombs could be made from available stocks of plutonium.

A second area for improvement was in the design of the explosive lenses. The crash program in February had abandoned several promising leads. Better designs could mean more power, and perhaps more importantly, more compact weapons. The explosive assembly itself was cumbersome, and Oppenheimer hoped that the lenses could be preassembled and shipped from Los Alamos in sealed containers. Both the electrical and fuzing systems were still marginal and could use better designs.

Oppenheimer also had to report that Los Alamos was still far away from knowing how to make a hydrogen bomb. Despite Teller's work, there were no firm leads on how to assemble a Super. The very lines of research in Los Alamos were concentrated on fission weapons. As such, they were too narrow for the enormous problems of the Super. Oppenheimer could do no more than recommend further study.

Oppenheimer's letter was hardly premature. Events at Los Alamos were moving with incredible speed. The gun was almost ready, and the implosion bomb was not far behind. The Laboratory's greatest challenge, the explosive lens, was on the verge of completion.

In Washington, Groves continued to push production schedules as much as he could. Los Alamos would need every bit of plutonium and uranium possible for its weapons. Stimson and his close advisers were trying to assess the bomb's potential for use in Japan and as a diplomatic weapon against the Russians. The new President was preparing for Potsdam and for what everyone told Truman would be "tough negotiations" with Stalin.

Everyone was waiting on Los Alamos.

PART TWO
OMEGA

**I AM BECOME DEATH,
THE SHATTERER OF WORLDS.**
The Bhagavad-Gita

13. HOMESTRETCH

Oppenheimer and his staff had had little doubt that a test of the implosion bomb was a necessity. If all went well, there would be slightly more than ten kilograms of Pu 239 available by late summer; that would be enough for two implosion bombs. Current production schedules called for enough U 235 for only one gun bomb by the same time. The implosion bomb was by far the more difficult weapon of the two and there were many questions about theory that only a test could answer. In a way, an experimental test of a bomb would also be a payoff to the scientists who created it; the bomb as a field weapon would be seen by a handful of Los Alamos men at best.

By March 1945, the war in Germany was nearing its end. On March 7, the American Third Army crossed the Rhine into Germany at Remagen. Within a month they would be at the Elbe. In the Pacific, the Americans were fighting a gruesome battle for Iwo Jima in which 4,000 men would die. On April 1, Marines invaded Okinawa to establish air bases that would put American bombers just over 300 miles from Japanese cities.

The German atomic bomb had been a phantom: Conversations with captured German atomic scientists revealed that endless quarrels among themselves had prevented a united effort. General Groves had eagerly pursued intelligence efforts since 1942 to garner information on the progress of German nuclear research. A special intelligence group called ALSOS was formed under the Army's G-2 section to conduct intelligence activities throughout Europe. The effort was led by Colonel Boris Pash and a scientist named Samuel Goudsmit. The new appointment redirected Pash's attention to the Germans and away from meddling in Oppenheimer's security history. Both men spent nearly two years darting in and out of German-occupied Europe collecting scientific intelligence.

Groves had prepared ALSOS for its task by asking Los Alamos and Manhattan scientists to assemble information that would assist agents without scientific backgrounds on what to look for in Germany and in occupied countries like France, Italy, and Norway.[1] While the information was scanty, it allowed ALSOS to confirm German activity. Nazi interest in a Norwegian heavy water (deuterium) plant led to a British commando raid that destroyed the hydroelectric facility for the rest of the war. Groves had been particularly concerned that the Germans would use radioactive materials to contaminate certain geographical areas, such as the Normandy beaches. But by March 1944, Groves was able to write General Eisenhower that while "radioactive materials" were known to the Germans, it was unlikely that they would be used. With the end of the war, Groves was pleased to see that his prediction had been correct.[2]

Hitler, whose imagination had always been captured by the spectacular rather than the strategic, had focused Germany's scientific resources behind the V-1 and V-2 rockets. He had never understood how the process of fission could produce a bomb, and had severely limited financial support to German physicists. As a result, the competition for funds had divided and crippled their efforts. Although it was not known to American intelligence efforts at the time, the Japanese were engaged in a very limited program of fission research. Before the end of the war, two Japanese scientists would be on the verge of a breakthrough in uranium separation.[3]

The Pacific war, however, had every promise of providing a bloodbath battle for Japan itself. For Stimson and Marshall, the battle for Japan was only a matter of time. Even as early as January 1944, Groves realized that the bomb would be developed too late for Germany. On January 17, he had been forced to write Field Marshal Sir John Dill of the Combined Chiefs of Staff to report that "the use of the TA (atomic) weapon is unlikely" against Germany.[4] By early 1945, the war in

Europe was nearly over. But Groves was not prepared to let the war in Japan conclude without the bomb's use. Anything less than a field drop of either Fat Man or Little Boy would diminish the importance of his Manhattan Project and lessen his historical legacy. He became more optimistic about the weapon's swift completion in Los Alamos; Stimson and his advisers became as hopeful. Roosevelt, until his death in April 1945, had seen the atomic bomb as more than a weapon. It was political leverage against the diplomatic tussle that was developing between the United States and the Soviet Union for control of the postwar world. Both Stimson and Marshall saw much the same advantage.

The Quebec Agreement signed secretly by Roosevelt and Churchill in August 1943 contained the explicit agreement that information on atomic weapons would be shared between the United States and the United Kingdom and no "third" party. With German control solid over Western Europe, only the Soviet Union possessed the ability to construct the huge separation plants necessary to manufacture uranium or plutonium, or to even employ bomb technology. The agreement, therefore, implicitly denied the Soviets access to American and British knowledge on the subject.

Both Stimson and Roosevelt agreed in December 1944 that the "secret" of the atomic bomb could be used as a means of obtaining a *quid pro quo* with the Russians.[5] The Combined Military Staffs of America and Great Britain were already planning global policies without involvement of the Soviets. While there was increasing pressure on Roosevelt from men like Niels Bohr and Leo Szilard to "open up" to the Russians, there was great resistance from within a small coterie of powerful American policy advisers to the President, as well as from British leaders. Stimson and others urged caution. Churchill's great distrust of the Soviets was well known to Roosevelt. While at Yalta in January 1945, the President tentatively suggested to Churchill that the two countries might brief Stalin on the work of the Manhattan Project. Churchill professed to be shocked and refused adamantly. He not only feared Soviet knowledge on the subject, but also feared that such information prematurely given would lessen Russian cooperation. Moreover, the Prime Minister feared that such sharing would also necessitate a similar revelation to the French.[6]

Roosevelt had nurtured a policy of cooperation with the Russians, beginning with Lend Lease and following with a series of concessions on military strategy. By the beginning of 1945, however, it was clear that Roosevelt and Churchill expected little cooperation from the Soviets. Both men closed off the possibility of wartime agreement for the control of atomic energy.

A stronger policy of restraint—and silence—was urged on

President Truman. After Roosevelt's death, Harry Truman suddenly found himself with a full disclosure of the promising work in Los Alamos. The spirit of "cooperation" toward the Russians fostered by Roosevelt faded quickly with Truman's quick entrenchment into the Presidency. His advisers were able to argue successfully against any further concessions to the Russians, particularly in the area of atomic bomb development. They contended that the possession of the atomic bomb would guarantee the United States a powerful advantage at war's end. Even Stimson, who thoroughly briefed Truman on the Manhattan Project developments, felt compelled to reflect that "we [the United States] must find some way of persuading Russia to play ball."[7] The Fat Man and Little Boy bombs were logical agents for pressure—if they worked.

The most immediate use of the bomb, of course, was as a weapon. The impending Allied assault on mainland Japan would be extraordinary in size and scope and would dwarf the landings in Africa and France. It would also be monstrous in cost: as many as 200,000 American lives might be lost. Groves felt that the atmosphere in Los Alamos might have become less pragmatic and too rarified. He increased pressure on Oppenheimer at every turn to complete both weapons as soon as possible.

There was still great doubt that the implosion bomb would work; theory said yes, but theory could be wrong. The Americans could hardly drop such an expensive weapon on the enemy only to have it fizzle. Oppenheimer met with his division and group leaders in March 1944 to discuss a test of the Fat Man. Everyone agreed that without such a test it would be impossible to understand certain conditions and effects of a nuclear explosion. Many theoretical gaps and uncertainties would remain, unless a full-scale test were made under experimental conditions.

There was unanimous agreement. Oppenheimer appointed Kistiakowsky to assume the leadership in planning for such a test, to be conducted in early or mid-1945. Kistiakowsky at the moment was deputy division leader for the implosion program within the Engineering Division. He quickly formed a new High Explosives Group under Kenneth Bainbridge to investigate and complete all the components for a test weapon: explosive lenses, tamper, firing circuits, and the like. In two months, after the August reorganization, Bainbridge's new group was given formal status and renamed X-2: Development, Engineering, and Tests.

Bainbridge went to work immediately. During early discussions of a test in the February Governing Board meeting, there had been talk of where such an explosion might take place. Because the possibility of a fizzle seemed so great, there were some discussions over methods of containing the initial explosion from standard explosives without losing the precious plutonium. Someone suggested that the bomb be detonated in a concrete tank, while another suggested either underwater or perhaps underground. After the meeting, Oppenheimer approached Bainbridge with the possibility of a separate organizational group within the Laboratory with total responsibility for a test. By late March Bainbridge was ready with preliminary plans.[8]

Bainbridge had to consider preparations for a test in the field in which they could study blast, earth shock, neutron and gamma radiations, and in which they could make a complete photographic record of the explosion and any atmospheric phenomena resulting from the blast. These basic experimental needs gave him the guidelines for organizing his staff.

Bainbridge quickly appointed Lewis Fussell, who had been working on detonators, to head a test measurements group. Donald Hornig was to cover earth and blast measurements. Herbert Anderson took on meteorological measurements and earth sample recoveries from the actual blast area. Phillip Moon, a member of the Laboratory's British team was in charge of measurements of nuclear process. And Julian Mack, also an Englishman, was to handle optical and photographic needs. Bainbridge suddenly had 25 men under him; in another year he would have 250.

During the early summer months, Bainbridge and Fussell prepared a systematic plan for a test shot scheduled in the first half of 1945. In a memorandum to Oppenheimer and others, both men outlined plans for a test based on a bomb that would deliver a force equivalent to between 200 and 10,000 tons of TNT. This plan was based on the assumption that a device like Jumbo would be used to contain the explosion.

Kistiakowsky read the memo and prepared one himself; he sent it to Oppenheimer with a copy to Groves. Plans for a test, he said, were based on the assumption that a nuclear explosion would take place. Such a test would provide information on the "various effects" of the gadget and would serve as the basic technical data for planning for both tactical use development in the future. The information would be useful in two ways:

First it will tell what is to be expected from a gadget of the same design as that to be tested. . . . Second, as the design of the gadget is improved and greater TNT equivalent is hoped for from subsequent gadgets, it will be possible . . . to predict with good accuracy the destructive effects from these improved models of the gadget.[9]

It was the sort of thing Groves liked to hear. It was also a subtle move to begin building justification for a test that would consume half of the world's supply of plutonium.

The code name Trinity came from Oppenheimer. Bainbridge soon realized that it was necessary to select a code name for the test site as a cover for the activity that was about to take place there. In September 1944, Bainbridge and one of his staff met with Oppenheimer in the Director's office in Building T. Bainbridge raised the question and Oppenheimer suggested Trinity after a few lines of John Donne's poetry he had read the night before:

> *Batter my heart, three-personed God, for you*
> *As yet but knock, breathe, shine, and seek to mend.*
> *That I may rise and stand, o'erthrow me and bend*
> *Your force to break, blow, burn, and make me new.*

It was only a casual suggestion but Trinity, or TR as it was to be called, came to describe a portion of harsh land in Southern New Mexico.

The choice of a suitable site had been an early concern for Bainbridge. He collected suggestions and assembled a list of eight different locations in the western part of the United States. The two best were the Tularosa Basin and the Jornada del Muerto in Southern New Mexico. The others were the military training grounds near Rice, California; the lava region south of Grants, New Mexico; an area southwest of Cuba, New Mexico; the sandbars off the coast of south Texas (now Padre Island), and the San Luis Valley region near the Great Sand Dunes National Monument in Colorado. A very simple requirement was that the prospective site be very flat, to minimize the effects of the explosion. Preferably, the location would have good weather, with light winds and as little dust and haze as possible. Ranches and small settlements must also be far away, to avoid the possible danger of radioactive fallout. The concept of "fallout" was new—less than two years old—and no one at Los Alamos thought it as more than a short-term problem.

Most sites were generally within driving distance from Los Alamos. Bainbridge thought that a flat, isolated site would be better for trucking in supplies and heavy equipment. California was too far. The San Luis Valley was inhabited by small groups of Indians, and Secretary of the Interior Ickes had recently forbidden the seizure of any lands belonging to Indians.

Bainbridge made auto trips to the regions north and south of Grants and Cuba, the Tularosa Basin, the Jornada del Muerto, and the California training grounds. Aerial surveys were taken, when possible, by either Bainbridge or Robert Henderson and his staff. Either Major Stevens or de Silva of Los Alamos Security accompanied them to represent military interests. The team decided to recommend the Jornada del Muerto.

This portion of New Mexico was a vast and unending desert that lay along the old Santa Fe trail between El Paso and Northern New Mexico. The name in Spanish meant "the dead man's route"[10] and was aptly coined: Spanish travelers from El Paso often encountered difficulties from Indians, or ran out of water, and many died.

To Los Alamos, however, the name was not yet ironic. Instead, they saw the desert as being well suited to their needs and having a major advantage: the land was already under War Department control. Hundreds of square miles had been converted to the Alamogordo Bombing Range, where B-17 and B-29 bombing crews practiced bombing runs before being sent overseas. The range took its name from the nearby town of Alamogordo, just a few miles north of the White Sands National Monument.

Bainbridge suggested to Oppenheimer and the Governing Board on June 29 that a final decision on the Jornado site be made as soon as possible. Oppenheimer preferred August 1 as a deadline. The next month would be used to obtain clearance and better topographical information. Groves approved the location in August and in September Bainbridge led a team to visit General Ent, Commander of the Second Air Force, whose headquarters were located in Colorado Springs. Bainbridge was accompanied by Parsons, Ramsey, and Major de Silva. Little information was exchanged between Ent and the Los Alamos team; no doubt they certified their request as part of an extraordinary operation. Bainbridge asked for an area approximately 18 by 24 miles within the northernmost part of the Bombing Range. Four alternative areas were discussed and Ent finally gave them the area $33°28'$ to $33°50'$ by $106°22'$ to $106°41'$. It was a good location. There was not a single town or ranch for 12 miles to the north and west and clear military land to the south. The nearest town was 27 miles away, and there was at least 18 miles of military reservation to the east to act as a cushion.

Fortunately for Bainbridge and the Laboratory, the Commander of the Alamogordo Bombing Range could provide them with extensive aerial photographs of the area. Bainbridge later recalled that "the Commander and his men flew that area and gave us detailed maps something like six inches to the mile where we could see every fencepost and prairie dog burrow in the place."[11] It was a break for Bainbridge; no adequate maps existed of the area and the photographs were invaluable. In June and July attempts had been made to obtain good maps, with only partial success. The ordering, of course, was done through the Security Office to minimize attention. Even while other sites were under consideration, Los Alamos ordered all the available U.S. Geologic Survey maps of New Mexico, Colorado, and California. Dummy agents were created to order county maps and grazing and state maps. Ent's aerial photographs turned out to be the most valuable source of information.

Once approval came from Groves, Bainbridge began planning for a central camp, scientific and technical buildings, and a series of roads to connect everything. A team was created, consisting of Bainbridge, Fussell, Captain Davalos of the Army, and the Post engineer. Together, they prepared a set of plans and submitted them to Oppenheimer and Kistiakowsky. Oppenheimer sent the package on to Groves on October 14 for his approval. Word came back on October 27 to proceed immediately. Contracts were let for construction and roads in early November.

Bainbridge gave the main living and work area the designation Base Camp. As it was completed in December, Groves ordered a detachment of military police to guard the place, and gave the assignment to a young lieutenant. Harold Bush arrived with an MP detachment on December 31, 1944. The Los Alamos security office sent down a "special agent," John Anderson, as the special representative of the Intelligence Officer at the Laboratory. Anderson took over the elaborate badge and pass system, the guard system, and communication between Trinity and the outside world.

Base Camp consisted of a few buildings for scientific work, a mess hall, several dormitory buildings, a meeting room, and a technical stockroom. The Camp was almost ten miles from the actual test spot called Ground Zero. There were three major observation and control shelters 10,000 yards from Zero, and located at each of three compass points. Each took the name of its orientation: North, West, and South shelters. To the east of Ground Zero lay the Oscuro Mountains; to the north was Santa Fe and Los Alamos; and to the southeast was Almagordo.

Most of Trinity was complete by early May. A complete communications system had been installed, including telephone lines, public address systems in all the buildings, and FM radios in a number of the jeeps and trucks assigned to the Trinity test. An elaborate stockroom, nicknamed FUBAR,* was stocked with everything from precision

---

*Short for "Fucked Up Beyond All Recognition."

instruments to toilet paper. Everything, of course, had to be trucked down from Los Alamos.

At the Laboratory, Bainbridge and his men continued to plan and schedule events. After the August reorganization, implosion became the major activity of the Laboratory; Bainbridge was able to draw men and resources almost as he needed them. Trinity grew by function. By March 1945, the TR organization had safety personnel, security, various military attachés and representatives, and scientific consultants. By June, the staff had grown tremendously, along with the planning. Fermi and Victor Weisskopf, for example, were asked to be chief consultants for physics experiments; Carlson and Joe Hirschfelder for bomb damage experiments. Norris Bradbury, who was later to become director of Los Alamos, was made responsible for the final assembly of Fat Man at Trinity. John Williams took on support services for the test. John Manley was given air blast and shock tests to design and run, and Julian Mack was asked to keep responsibility for spectographic and photographic records of the test. Other men were given responsibility for health and safety and meteorology.

In mid-spring, Frank Oppenheimer arrived from Oak Ridge to assist his brother. General Groves had sensed the tension pressing on Robert Oppenheimer and asked both Frank and Isidor Rabi—two of the men closest to Oppenheimer—to come to Los Alamos to lend support to the hard-pressed director. Frank assumed an active role at Trinity as special assistant to Bainbridge. For most of June and early July, Frank made the rounds at Trinity in the oppressive heat, checking for bottlenecks or potential trouble spots in the elaborate series of experiments designed around the first test of Fat Man. Frank fitted comfortably into the ménage of desert scientists. He made endless checks of equipment and experiments for Bainbridge and offered a hand and moral support where he could. Rabi arrived just days before the test. He stayed close to Robert Oppenheimer and injected a calculated calm into the frenetic atmosphere.[12]

Bainbridge originally planned the Trinity test with the belief that a misfire was a strong possibility. In late 1944, alternatives for containment or baffling were taken seriously. At the June 6, 1944, meeting of the Governing Board, Bainbridge reported on these concerns, as well as on some experiments that had been conducted to determine whether active material could be recovered from a misfire. Recovery from sand would be possible, he reported, but it would seriously restrict the collection of vital experimental data. In July, Jim Tuck suggested to Oppenheimer and Bainbridge that exploding the bomb in a large wooden tank might work. Such a tank, he suggested, could be placed in a shallow gully with its surface macadamized or at least waterproofed. Unfortunately, firing the weapon in a hole in the ground was not satisfactory because recent experiments showed that such an explosion would confuse instrumentation above the ground.[13] Kistiakowsky and Neddermeyer still favored a large pressure vessel; Oppenheimer tended to agree. Fortunately, confidence in implosion grew, and in March 1945 all plans for Jumbo or similar containers were dropped.

The concern over collecting test data was not unrealistic. Purportedly, the major justification for a test lay in proving the weapon a success *and* in collecting data to fill in between theory and fact. Many planning months were given over to determining the appropriate experiments to run, scheduling their logistical needs, and obtaining and transporting the necessary equipment and instrumentation to Trinity. New gauges were devised to measure the blast; geophones were selected and laid to determine earth shock; thousands of miles of cable were laid to connect instruments with control panels and communication lines.

There was already some idea of the potential explosive force. The best guess at the time seemed to suggest something around 10,000 tons of TNT. Even some of the effects of the bomb were projected. In March 1944, Hans Bethe and Robert Christy prepared a memorandum on the "Immediate Aftereffects of the Gadget."[14] The immediate area around the explosion would reach a temperature of about one million degrees and would have a radius of about 30 feet. In less than 1/100 of a second, the central fireball would expand to about 800 feet in diameter and cool down to 15,000 degrees. Shortly thereafter, the fireball would rise into the stratosphere and in two or three minutes would be at an altitude of 15 kilometers with temperatures of about 8,000 degrees.

Bethe and Christy warned that reconnaissance planes in the area would have to be careful. In the first tenth of a second after detonation, a flash of bluish light was to be expected. It would be as bright as the sun at a distance of about 100 kilometers. No plane should be any closer than seven miles from Ground Zero, and personnel should not look at the explosion until after the first brilliant flash of light. The danger from radioactivity was low, Bethe added dryly, because of the distance of the plane from the actual source of neutrons.

Bainbridge spent most of his time preparing schedules. Weekly meetings were held in which personnel from other Laboratory groups and divisions were invited to propose new experiments or to specify logistical needs. Bainbridge made a point of reporting Trinity progress at each meeting. This was the most likely method, he thought, of determining where delays were likely to occur. Those scientists wishing to conduct experiments were asked to prepare full proposals detailing objectives and methodology, as well as expected personnel, equipment, and time needs.

A number of dates were set for the test and were subsequently canceled. In March 1945, a test date of July 4 was set and very soon was found to be unrealistic. From an outside contractor there was a delay in the delivery of full-scale explosive lens molds, and therefore a delay at Los Alamos in producing shaped explosives. There was a delay in the shipment of plutonium from Hanford. Bainbridge reported this to Oppenheimer and the Cowpuncher Committee. With their overall command of Laboratory scheduling, they attempted to place priorities within the Laboratory to coincide with outside deliveries. By the middle of June, the Cowpunchers set July 13 as the earliest possible date for a test with July 23 as a more probable date.

The committee kept one particular individual busy with their jostling of dates and times. John Hubbard tried desperately to keep on top of what scientists considered favorable weather for their experiments. Hubbard was the chief meteorologist for the Manhattan District, and was placed on assignment to Trinity by Groves. Hubbard surveyed every unit within the Laboratory and tried to find a time that best suited everyone's requirements. Drawing on the Army Air Forces Weather Division, he had information from worldwide sources. He ultimately recommended to the Cowpuncher Committee that July 18 and 19 was a first choice; 20 and 21 and 12 to 14 a second choice; and July 16 as a third choice. July 16 was only a probable date, he insisted.

With this information the Cowpuncher Committee met on June 30 and reset the July 13 date for July 16, 1945. As it turned out, this date was the earliest possible time in which all components of the Fat Man would be available to Los Alamos. Oppenheimer had already made a commitment to Groves to have the test as soon after July 15 as possible. President Truman was expected to be in Potsdam at the time and needed all the political leverage possible for negotiating with the Russians. The President had only weeks before been given a full account of the Manhattan Project and the weapons nearing completion at Los Alamos.

General Groves had set a completion date of July 14 for the Trinity test. On the morning of July 2, Oppenheimer telephoned Groves in Washington and requested a delay of three days. Fat Man would be ready certainly by July 17. Groves refused, citing the importance of the Potsdam Conference. The request seemed so earnest, however, that Groves called back in the afternoon to reaffirm the July 14 date; his superiors, he said, were adamant about the fourteenth.

At Base Camp work became frenetic. The weather during

the early summer was very hot, with daytime temperatures often soaring over 100 degrees. Any delay in procurement or delivery meant that Trinity teams had to work doubly hard when the equipment finally arrived. By late March, the normal working day of ten hours was extended to eighteen hours. The Services Group under John Williams was responsible for the thankless task of providing the necessary wiring, power, transportation, communication, and construction in face of the increasing daytime temperatures.

For a month before the test, the TR teams based at the site met nightly to hear reports on construction, air complaints, and process requests for the following day. Bainbridge had to weigh each request and make personnel assignments for the following day. Construction men were placed on the highest priority.

Groves was insistent that the work at Trinity not be connected openly with the work at Los Alamos. The site's location on a military reservation was most helpful, but precautions were taken whenever possible. Bainbridge, in a memo written on March 14, issued a list of regulations. Individuals or groups leaving Los Alamos for Trinity were made to check in at Major de Silva's office. Everyone at Trinity was required to sleep and eat at the campsite and not in nearby towns. Not only the military, but civilians as well, were required to spend all their time at the site and not in towns at movies or dinner.

A week later, on March 21, Bainbridge issued another memo with instructions governing the drive from Los Alamos to Trinity:

The following directions are strictly confidential. . . . Under no conditions on this trip, when you are south of Albuquerque, are you to disclose that you are in any way connected with Santa Fe. If you are stopped for any reason and you have to give out information state that you are employed by the Engineers in Albuquerque. Under no circumstances are telephone calls or stops for gasoline to be made between Albuquerque and your destination.[15]

The one approved stop was Roy's Café in Belen. It was everyone's suspicion that Groves had placed one of his G-2 security men as chef there. There were pressing exceptions: hot travelers reaching San Antonio in the late afternoon often stopped—illegally—for a beer at Jose Miera's bar.

There was a very thorough pass system. Departing men from Los Alamos secured a pass from Bainbridge's office and presented it at the base of the first guard tower at Trinity. The pass was exchanged for a badge. The badge was worn at all times and surrendered upon leaving

Trinity, when the original Bainbridge pass would be returned. The military police kept careful rosters at all times, down to the number of trips made to Trinity and the amount of time spent there.

Security accidents occurred despite the heavy precautions. Construction workers from Los Alamos somehow found their way down to Trinity and recognized some of the scientists there. As construction began at Base Camp, it was evident that two-way radios would be necessary. Guards, for example, needed to maintain communication among themselves for security checks, and when the test occurred, it would be necessary to have ground-to-air communication with the scouting planes.

A request was made of Washington for special radio frequencies to be assigned to Trinity that could not be monitored. After months of waiting, the assignments came back. Strangely enough, the ground frequency assigned to Trinity was the same one assigned to a railroad freight yard in San Antonio, Texas. The ground-to-air frequency was the same one assigned to the Voice of America. Bainbridge felt quite sure that each could hear the other.

Safety was a second concern. What if radioactive dust drifted over nearby towns? Plans were made for Major T. O. Palmer of the U.S. Army to be stationed north of the test area with 160 enlisted men on horses and in jeeps. Palmer was instructed to evacuate ranches and towns at the last moment if necessary. Another 20 men in Military Intelligence were disguised as civilians and stationed in nearby towns and cities up to 100 miles away. Most of these men were armed with recording barographs to get permanent records of blast and earth shocks. The nearest towns were the most obvious candidates for disaster: San Marcial, San Antonio, Soccorro, Carrizozo, Oscuro, Three Rivers, Tularosa, and Alamogordo.

Groves was aware of potential legal complications, and indicated particular concern over damage to houses and buildings; danger from radioactive fallout was second. Bainbridge reported these concerns to his key personnel on May 2.[16] As a result, elaborate plans were drawn up for the test date in the event that it would be necessary to evacuate personnel. The Laboratory's medical officer, Dr. Hempelmann, was placed in charge.

Bainbridge was not unaware of the possibility that the "best laid plans" might go awry. Even as early as the summer of 1944, there were recommendations from some of his staff that a rehearsal shot be made with conventional explosives. The test shot would be useful in testing procedures and in providing training for an implosion shot.

Moreover, such a pretest would provide a means for calibrating blast and earth shock equipment. Oppenheimer readily agreed that a preliminary test would be useful. A tentative date in May was set.

Little information existed on the effects of large explosions. Los Alamos calculated that 100 tons of TNT would be a base amount. A few miles from Ground Zero at Trinity a site was prepared. Workers constructed a 20-foot tower from railroad ties and lumber and stacked 100 tons of explosive that were hauled by train from Fort Wingate. Fortunately, the explosives could not be detonated by shock, because several cartons of TNT were accidentally dropped.

Scattered within the boxes of TNT were 1,000 curies of radioactive fission products from the Hanford Plant. Each radioactive slug was sealed in a plastic tube. The purpose of the material was to simulate at a safe level the radioactive products that would be spread from a nuclear explosion. The TNT added another 18 feet to the wooden platform. The platform and explosives were carefully arranged to approximate the 100-foot tower that was being built for Fat Man. By using scaling techniques, the scientists could calculate the effects of the Trinity test, and measurement instruments could be calibrated. Photographic equipment was placed at distances from the platform that matched the distance from the Trinity tower.

Hubbard told Bainbridge that his meteorological data suggested May 7 at 4 A.M. would provide ideal weather for the test. With everyone in place, and recording instruments running, the test had to be stalled for thirty minutes while the observation plane maneuvered into place. At 4:38 A.M. the TNT was ignited from several locations within the stack. An enormous and highly luminous orange sphere appeared and quickly dissipated into an oval; within a few seconds it bloomed into a mushroom shape and rose 15,000 feet.

The test was extremely successful. Bainbridge confirmed that his staff expectations for Trinity were accurate, and that test data provided a base for experiments next month for Fat Man. There were failures, of course. Men forgot to start two cameras at the north and south stations, and another man forgot to release several flash bombs. Two other men with experiments to run were forced to spend the night in the desert when their jeep broke down. Generally, the explosion gave everyone an insight into the handling of their equipment under desert conditions.[17]

The successful rehearsal lifted everyone's spirits. Bainbridge's planning for Trinity was essentially on track. Daily problems continued: procurement, personnel shortages, the heat and dryness, the general exhaustion of working 16 and 18 hours a day. Procurement was a major headache. Equipment ranging from pencils to heavy equipment was needed urgently and intact in the middle of the desert. Robert Van Gemert headed procurement and found himself snarled in supply demands that all had top priority ratings.

A year earlier, the Laboratory had been forced to devise its own rating system for procurement; four ratings were created. These four—X (the highest), A, B, and C—became meaningless as Trinity neared. By May 1945, the X rating accompanied all requests, and Van Gemert was again forced to subdivide the system: XX, X1 and X2. The XX rating was to be used only if failure to obtain the needed material would produce a major setback to the overall program of the Laboratory. The rating system authorized the Laboratory's Procurement Office—through the Washington Liaison Office—to use its recourse to the highest authorities of the War Production Board and all government agencies. In these cases, special cargo planes could be dispatched from anywhere in the United States to get delivery.

Van Gemert also suffered from the poor communication between Trinity and Los Alamos. Communication was limited even more by the use of major purchasing agents in Los Angeles and other key American cities. The cardinal rule was that there was to be no direct communication between Los Alamos and suppliers. Scientists were forced to make the trip between the test site and the Laboratory, or to send a representative. They met delays on the Hill, and procurement was snarled there by poor service from business and industry. Everyone in the military had his own agent who regularly pounded desks in civilian businesses. Most government agents were happy to obtain a 15-week delivery time. Los Alamos asked for a three-week turn-around time.

Some items were impossible to get at first try. Seismographs, for example, were needed to test earth shock waves in areas around Trinity. The only instruments available had been sold a week before to the Argentine government. Only through the direct intervention of Groves and his office was the Laboratory able to reverse the sale and instead have the items sent to Trinity. Ordinary garden hose was urgently needed to protect sensitive wires between instruments and control points. An order for 10,000 feet was lost during a shipping strike and substitutions had to be found.

Oppenheimer finally had to call a meeting in May to discuss the backlog of urgent requests for equipment and supplies. A review of the situation revealed that one problem was the shortage of personnel in the Los Angeles, Chicago, and New York Manhattan Project purchasing offices. Although requests from all Manhattan projects had grown tremendously in two years, the number of people at each site had remained the same. Groves was consulted. Pay raises were given, and new staff members were hired. At last Groves permitted direct communication between Los Alamos and New York and Chicago.

Service improved, but hardly in time for Trinity. Procurement lags continued until the day of the test. Everyone simply had to do the best he could with what was available. Communication never improved, and only five people in the Laboratory were authorized to telephone between Trinity and Los Alamos. Even then, the calls were routed through Denver to Albuquerque, a small town called San Antonio, and finally to Trinity. Most communication had to be by note. Courier service was created with at least two, and often more, trucks leaving Los Alamos every night to avoid both the heat and the attention. The trucks were often told to stop at the offices of the U.S. Engineers in Albuquerque to pick up odd packages addressed to Professor W. E. Burke of the Physics Department at the University of New Mexico. Burke was a fiction created by the Laboratory security men further to avoid connections between Los Alamos and small suppliers across the country.

Methods of delivery varied and occasionally bore touches of elegance. Once, 24 rolls of urgently needed recording paper made their way from Chicago to Albuquerque hidden in a first-class Super Chief drawing room. On another occasion, a trainload of telephone poles for use at Trinity was needed, but no freight train could make the railroad siding near Polk, New Mexico. After some friendly persuasion, a carload of such poles was attached to the rear of a passenger special and delivered by the Texas Chief.

These were light moments, but for the most part, the desert at Trinity was relentless. Sanitary conditions were hard to maintain at Base Camp, particularly in the Mess Hall; the water from nearby wells was so hard that it quickly clogged all pipes. Scorpions, snakes, and other desert creatures were a constant menace. The SOP (standard operating procedure) of the day was to begin each morning by carefully shaking out one's clothes before dressing.

Everyone had to become inventive to avoid the insects and to amuse himself. A nearby abandoned ranch house had a water reservoir that was used daily for swimming. Passing herds of antelope contributed to the camp's menu through the use of the Security Patrol's submachine guns, and wandering cattle occasionally found themselves on the menu as well. A beer fund was maintained by staff as protection against dehydration and because nearby towns were off limits. Everyone was certain that the Army owned only five movies and showed them every night for nearly six months. Before the accelerated schedule in late June and July, the day at Trinity began at five, with breakfast before six, and then off to the day's work. Dinner came at six in the evening, and after-hours were largely spent in late work or informal briefings. John Williams, among other group leaders, held nightly meetings every day for a month before the test.

Under the excitement and exhaustion there was still a sense of uncertainty: Would all their work be in vain? Doubt displayed itself in many forms. For some, it meant trying to double their efforts through checking and rechecking. For others, it emerged in the form of black humor, such as the parody that circulated the Laboratory in July:

> *From this crude lab that spawned a dud*
> *Their necks to Truman's axe uncurled*
> *Lo, the embattled savants stood*
> *And fired the flop heard around the world.*

And so it went. The days grew longer and hotter, the work more exacting and demanding, the urgency for completion and perfection more intense. Everybody knew that the "day" was drawing near and that somehow his work and sweat was going to be part of it. Oppenheimer was restless and apprehensive. His staff and his division and group leaders were apprehensive. Los Alamos began its summer. There was a tension, subtle at first, that grew until wives could tell something was near.

Groves organized a trip to the West Coast to visit Manhattan projects there. Secretly, he wanted to be nearby when the day of the test came. President Harry S Truman prepared for his first direct meeting with Stalin. War Secretary Stimson had already alerted him that he might have an ace card to play at Potsdam. From the East Coast, Tolman, Bush, Conant, and Compton made clandestine preparations to attend; from the West Coast would come Lawrence and Groves. All were coming for the sole purpose of bearing witness to the birth of a new age. Latitude 33°40'31" and Longitude 106°28'29" was soon to open that age.

14. DAWN

Dead Man's Route: no one at Trinity disagreed on the appropriateness of the Spanish name. Only the huge, rambling Oscuro Mountains to the east of them broke the harsh and monotonous landscape. Most of the men from the Hill came down Highway 64 from Albuquerque through the small New Mexico towns of Belen, Lemitar, Socorro, and San Antonio. Turning left in San Antonio and driving east toward Carrizozo, they came directly into the trailing path of the Jornado del Muerto. Almost in the middle of the etiolated valley, the tired drivers turned right for a drive that put them between the Oscuro Mountains and several smaller mountains in the distance to the west.

Here was true desert landscape: few trees of any size and scattered clusters of desiccated shrub. The road to Trinity was largely dirt, and the cars and trucks threw up clouds of loose arid dirt as they clattered down the rutted road to Base Camp. The flatness of the terrain was broken only by small risings and fallings of the elevation of the road as it wore its way to the center of the valley. It was impossible to see much of anything from the road, especially when behind another vehicle, until one was almost on it. Base Camp suddenly appeared, preceded only by Army jeeps with long FM antennas. In the distance was a tall metal tower.

The site was spread out over several dozen square miles, with the center of human activity at Base Camp. There was a constant flow of men and trucks in and out of the Camp, depositing supplies or picking them up, and coming to eat or sleep and check schedules or solve problems. Everyone's attention focused on a central hub: Ground Zero.

Williams and his men had completed Bainbridge's request for a 100-foot tower with a small three-walled building at the top. Radiating out, like long thin arms, were three straight roads which Groves had finally agreed to have blacktopped at a cost of $5,000 a mile. Ten thousand yards from Ground Zero along each road was the first of three shelters constructed of wood and concrete and covered heavily with earth. Each shelter, with a senior scientist in charge, had major functions to perform during the test. Each shelter was called by its direction: North 10,000 and so on.

As soon as the test occurred, a medical officer would take charge in the event it was necessary to evacuate. Each shelter had been issued a full complement of radiation-detection devices and enough vehicles to evacuate each person assigned there. Robert Wilson was assigned North 10,000; John Manley was given West 10,000; and Frank Oppenheimer was assigned South 10,000.

The date of final countdown was announced on July 1 and circulated in a memo two days later. Rehearsals would be held on July 11, 12, 13, and 14. On July 5, Oppenheimer cabled Compton in Chicago and Lawrence in Berkeley:

Anytime after the 15th would be a good time for our fishing trip. Because we are not certain of the weather we may be delayed several days. As we do not have enough sleeping bags to go around, we ask you please do not bring anyone with you.[1]

The Metallurgical Division was working long hours to finish the two plutonium hemispheres to be used in the test. The two half-spheres had to accommodate the small polonium neutron source, as well as fit perfectly within the uranium tamper shell. Close fittings were absolutely necessary, and any aberrations in the surface had to be corrected. A last-minute polishing job to repair surface blisters was apparently successful. On July 11, in a convoy led by Army Lieutenant Vaughn Richardson, the plutonium made its way to Trinity in the back seat of a car. When the cars arrived at Trinity, the young officer sought out Bainbridge at the base of the tower and asked for a receipt.

Bainbridge was surprised and rankled. It was a waste of time. He directed the Army men and the plutonium to the nearby McDonald ranch house used as a makeshift laboratory for the final assembly of the bomb's core. The young military men seemed very eager to unload their cargo, even though they weren't supposed to know what the small box contained. General Thomas Farrell, the special assistant to Groves and deputy to the Manhattan District, finally arrived and provided his signature. With the receipt signed, the plutonium was turned over to Louis Slotin.

Norris Bradbury, who was in charge of the bomb's high explosives, began his own countdown in Los Alamos on July 7. On that day he put the Fat Man's explosive lenses through a number of tests to study methods of loading and unloading for shipment and to determine whether transportation would cause them any damage. Even a slight flaw or crack might cause an irregularity in the simultaneity of explosion. On July 10, the crews in Los Alamos began a twenty-four-hour-a-day job of preparing the lenses for shipment to Trinity.

The intricate fitting of the explosive lenses raised questions about the gadget's ability to make the trip from Los Alamos down to Trinity. On July 3, a dummy bomb without the plutonium core was assembled and shipped in a truck to Trinity. Once there, the dummy was allowed to sit at the base of the tower for a day covered by a tarpaulin. The next day, the gadget was driven back to the Laboratory for a thorough check of all parts. Fat Man had survived the trip.

An earlier concern had been that the explosive lenses might accidentally explode under combat conditions if hit by enemy antiaircraft fire. At Los Alamos Bainbridge designed a test that involved a full-scale Fat Man with a dummy core. From a distance, the bomb was fired at with 20-millimeter cannon shells, such as the Japanese might use against an airplane. The Fat Man did not explode, and the test suggested that unless a detonator was hit directly, the bomb would have a fair chance of surviving enemy fire.

Both Kistiakowsky and Bradbury personally examined each lens at S Site laboratories. They checked for cracks, chipped corners, and other imperfections that might somehow affect the explosion. Each lens was then X-rayed and assigned a serial number. After all lenses passed the inspection, they were reassembled and encased within a metal shield composed of pentagonal pieces that bolted together to form a sphere. At one o'clock in the morning on July 13, the preassembled explosives began their final journey to Trinity in another convoy of Army trucks led by Kistiakowsky and Lieutenant Richardson.

Twice in May, airplanes from the nearby Alamogordo Air Base strayed over Trinity. Although air traffic was banned for the Trinity area by the Base Commander, several planes nevertheless wandered from their nighttime course and mistook the lights of Base Camp for their illuminated targets. They dropped bombs which hit the carpentry shop and the stables, and started a number of small fires, but the Camp suffered little damage and no one was hurt. The day before the test all bombing runs were canceled at Alamogordo. Pilots and crews were angered. The Air Base was the final stop in their training before they were to be sent overseas, and most crews needed as many flight hours as they could get. The ban was scheduled to lift shortly after the test, and Groves learned later that many crews were already on the airstrips when the test occurred.

The test rehearsal conducted on May 7 had shown the importance of careful planning and double-checking. For Trinity there would be three major experimental programs. One of the most important tasks was to determine the nature of the implosion process. Intervals between the firing of the first and the last detonator—there were thirty-two in all—would be studied by Kenneth Greisen and Ernest Titterton to give the scientists a better understanding of simultaneity. Robert Wilson and Darol Froman would try to determine the time between the firings of the detonators and the emission of gamma rays. This information would give key evidence on the behavior of implosion. Bruno Rossi was to study the fission rate.

Emilio Segré took on the job of studying the effectiveness of the bomb in its release of nuclear energy. To do this he had to study gamma rays in the fission products released. His colleague, Hugh Richards, was to do the same thing with delayed neutrons. Herb Anderson would enter Ground Zero shortly after the explosion and capture soil samples to determine the ratio of fission products to the amount of unconverted plutonium. The combined results of these experiments would give the Laboratory a patchwork portrait of implosion processes.

John Manley was taken from his Cockcroft-Walton generators and asked to conduct damage and blast measurements. Manley was surprised at the assignment, and indicated that he knew nothing about conducting experiments on blast effects. He was told by Oppenheimer that there were no experts. William Penney was given the task of measuring the heat given off by materials that were ignited by the explosion. Julian Mack presided over a variety of optical and photographic experiments that would measure and record visual effects, such as the fireball. Mack used both color and black-and-white films and movie cameras that ran from below normal speed to over 8,000 frames per second. Observation planes manned by Parsons and Luis Alvarez would be making photographic records from the air.

Herb Anderson had been assigned the task of testing the soil around Zero immediately after the test. A number of schemes were proposed, including the use of a semirigid blimp with an ingenious scoop, that would descend as soon after the blast as possible and collect a sample. It was feared, however, that the desert temperatures would make the blimp unwieldy. Even helicopters were considered. Two Army tanks were finally chosen as recovery vehicles and were shipped to Trinity sealed with lead linings.

Those concerned with the bomb itself worked in the heat at Ground Zero. The 100-foot tower had a small covered building on top; the ladder was located on the northwest corner. On the west side of the tower, jutting out from the open wall, was a small balcony about four feet wide and running the full length of the tower. A lift was constructed to bring the bomb up through a three-foot trapdoor in the floor of the platform. The 18-by-18-foot base of the tower was concrete, and Groves had ordered the surrounding dirt to be covered with asphalt to minimize dust. A tent had been hastily constructed around the base of the tower to shield the bomb, or "gadget" as it was called, from the wind and loose dust.

The Trinity area had once belonged to a family of ranchers named McDonald. Their stone house had been commandeered for the preparation of the plutonium sphere, initiator, and tamper. The house had four rooms and a cellar. Two rooms were selected for assembly work and were thoroughly cleaned, the walls and floor scrubbed, and the windows sealed shut. From the inside, the window frame was completely covered with flexible plastic sheets nailed to the wooden frame. The edges of the sheet were then sealed with masking tape one inch wide and then again with tape three inches wide. The same three-inch tape was placed between the floors and walls, walls and ceiling, and on all corners. It was hoped that fastidiously sealing the rooms would prevent the possibility of introducing dirt into the weapon.

Marshall Holloway was responsible for pit assembly, the final merging of the fissionable material and the high explosives. Earlier in the day, Holloway had been assisted by Robert Bacher and three others at the McDonald ranch house. In one of the sterile rooms, eight men with white gowns and gloves had used a table covered with ordinary brown wrapping paper. They carefully laid the plutonium hemispheres down and gave them a careful inspection. The plutonium was slightly warm, as was the initiator. At midafternoon, the initiator and plutonium—called the "plug"—were slowly driven by car to Ground Zero.

At one o'clock, a team of men gathered at the base of the tower to assemble the weapon for the last time. Norris Bradbury led the assembly team, which included Kistiakowsky and six other men. The delivery truck from Los Alamos was backed up to the base of the tower and the tarpaulin removed. Some of the men were now getting their first look at the complete bomb. A hoist carefully lifted the dark sphere off the truck and lowered it to a wooden cradle directly below the trapdoor 100 feet up. The metal sphere was designed with a polar cap that could be removed along with a dummy plug. This section of the bomb was left facing up. A specially fitted polar cap replaced the dummy used in shipping, and the new cap had been built to contain a funnel. It was now three o'clock and the heat from the afternoon sun was terrific. Bradbury and Kistiakowsky waited fifteen minutes for a general inspection to be performed and then turned the bomb over to the Gadget Group for an inspection of the plutonium and initiator.

Holloway, Bradbury, and members of G Team carefully began placing the plutonium sphere into the bomb. With the gadget's polar cap resting up, the hoist was used to lower a cylindrical plug into the bowel of the bomb, with a pair of tongs guiding the way. A problem suddenly confronted the men: the plug would not slip into place, but seemed lodged against the uranium tamper. Every piece of the bomb's core had been individually machined in Los Alamos to fit within a few thousandths of an inch; this was necessary to maximize the effect of the uranium tamper. Holloway immediately suspected that the desert heat had expanded the hemisphere. He checked with Roy Thompson, who stood nearby. Both men agreed to wait a few minutes to see whether the plug would settle down. When Holloway looked again, he found that the plug had sunk neatly into place. With relief the men realized that the plutonium core had expanded during the trip down from Los Alamos. As the two pieces rested against each other, the heat was exchanged from one part to the other, permitting the two to slip into place.

Just after six o'clock, Bradbury's Explosive Group assumed command again and inserted the remaining explosive lenses with a small vacuum cup that facilitated lowering the charges. The men used small pieces of paper where the fit was not precise and were prepared to use a small hypodermic needle grease gun to facilitate putting the parts together.

Shortly before the last metal pentagonal shield was attached, Boyce McDaniel inserted a long, thin-walled hypodermic tube with a magnesium wire running thorugh it. This wire could be withdrawn at regular intervals and checked with geiger counters for any abnormal rise in radioactivity. With the last shield attached, the dull-black gadget was again turned upright and left surrounded with the tent. The bomb was left under MP guard until Saturday morning.

At eight o'clock the next morning, the gadget was slowly lifted to the top of the tower through the open trapdoor. The floor was reassembled, and the bomb lowered to a special cradle within the small building. At nine o'clock, a team headed by Greisen climbed the tower to begin attaching the detonators and the necessary electrical cables. All five pairs of electrical leads and one coaxial lead to the gadget were disconnected. Greisen gathered them together and took them down from the tower to be kept safely stored until the arming party used them at night. Another series of leads was taken from a black box known as the informer, and similarly stored. Once the cables were removed, there could be no further tests of the detonators or the electrical system. At the same time, there was no possible way to detonate the bomb by accident. Thirteen and a half pounds of plutonium sat quietly surrounded by 5,000 pounds of high explosives.

The metal gadget was grounded electrically until the last trip from the tower. Both Bradbury and Kistiakowsky journeyed to the top to verify that the leads were disconnected. Bradbury's thorough "hot run" schedule concluded with two items, "Sunday the 15th: Look for rabbit's feet and four leaf clovers." "Monday the 16th: BANG!"

Sunday moved slowly. The day's work called for checking and rechecking the hundreds of instruments and the circuits that fed them. Joe McKibben had the trying job of supplying the timing and remote operating signals that controlled the experiments. For two weeks he and his staff had been badgered by each scientist whose equipment required one of McKibben's control wires. Experimental equipment was scattered across the desert and hundreds of miles of wire had to be laid and checked.

Oppenheimer was very tired and nervous, fretting and moving from one location to another. A few months earlier, Bainbridge had been asked by Richard Tolman and General Farrell to keep Oppenheimer away from the final assembly of Fat Man and the tower at Zero. Apparently under urging from General Groves, they had feared that Oppenheimer's state of mind might be affected by the great tension and pressures surrounding the test. Bainbridge immediately balked. The bomb belonged to Oppenheimer more than anyone else, and under no circumstances could he "forbid" his director's presence during any phase of the test.[2]

Groves arrived Sunday afternoon, along with Bush and Conant. Several busloads of scientists were on their way from Los Alamos to view the test from Compañia Hill, some twenty miles to the northwest of Ground Zero. Cars were dispatched to Santa Fe and Albuquerque to pick up the notables. Among them were Charles Thomas, the Manhattan District's coordinator for chemical research, Ernest Lawrence, Sir James Chadwick, and William Lawrence of *The New York Times*,* the only newsman allowed to cover the test. Others, such as Tolman, C. C. Lauritsen, Isidor Rabi, Sir Jeffrey Taylor, and John von Neumann, were expected later during the day.

Instructions were issued to those watching the test. Everyone was urged to take a position on Compañia Hill or at Base Camp. For those at the Camp, a warning signal consisting of a short blast would be given at five minutes before the test. At that time everyone was to face south, opposite Ground Zero. Three minutes later a long siren would sound, and people were expected to sit on the ground or in a shallow depression with their faces and eyes directed south. After the initial explosion and burst of light, a person could look toward Zero through welders' goggles or a piece of special glass. The two-by-four-inch rectangle had been given out in an envelope certifying that the glass met federal specifications for use in arc welding helmets. Everyone was instructed to remain in position until the blast wave passed over, about fifty seconds later. Two short blasts would then signal the okay to rise. There were similar instructions for the excited men on Compañia Hill.

There was also high interest in Los Alamos. Many men who couldn't be at Trinity or on Compañia Hill decided to try for a view of the shot from the roads around Trinity or from the Sandia Mountains outside Los Alamos. The Laboratory's Intelligence Office caught wind of these plans and tried to discourage such trips. Security precautions were accelerated just before the test. All telephone conversations were secretly monitored and agents in Santa Fe and other small towns were alerted to catch security infractions. One scientist was caught at the La Fonda Hotel in Santa Fe telling a companion that a drive to Socorro about four in the morning would bring a sight that the man would never see again. Unknown to the scientist, the companion happened to be a confidential informant and turned in the scientist's name to the Security Office.[3] The man eventually resigned.

---

*Lawrence's eyewitness report was not published in *The New York Times* until after news of Hiroshima was made public by President Truman.

On Sunday morning Military Police at each gate to Los Alamos began to record mileage readings from all automobiles leaving the Hill. These notes were compared against returning mileage counts and at least ten men were suspected of having traveled far enough to view the explosion. A number of individuals, including a few wives, were more devious. They drove or hiked only a few miles away from Los Alamos to mountain tops, where they had relatively unobstructed views of Southern New Mexico. The bomb, after all, was the baby of the Laboratory, and there was little the Security Office could do to dampen parental interests.

At Trinity, tension and excitement grew throughout the day. Only the weather seemed unconcerned. By Sunday evening the skies grew dark and thunder began to rumble in the distant mountains, rolling its sound down to Trinity. Lightning cracked, and it began to rain. Alone in the tower's cabin, Donald Hornig huddled against one wall with Fat Man only a few feet away. He had been reading an adventure novel set in the South Seas, but now he waited apprehensively for the storm to pass. Many spectators and scientists moved into the shelters or into the buildings of Base Camp. Some of the hardier souls remained outside. Those on Compañia Hill took shelter as best they could, in cars or under covers, and shivered in the cold night air.

Shortly after eleven o'clock, Bainbridge gathered together the arming party. He, Kistiakowsky, and McKibben rode in one car; a weather group, consisting of Jack Hubbard and two Army sergeants, clustered in another. Driving a third car, probably under orders from Groves, were Lieutenant Bush and another MP. On the way to the tower, Bainbridge stopped at South 10,000 Shelter and locked the main arming switches. Donald Hornig was already preparing to connect the firing circuit from the squat dummy control used as an equipment check to the real unit. A final check was made of the firing connections that ran to the sleeping bomb.

After the check, Hornig prepared to move to South Shelter, where he was responsible for the STOP switch. If anything went awry with the automatic firing devices, he would flip the STOP switch and close down the firing circuit to prevent an explosion. Kistiakowsky joined Hornig in the tower and descended to adjust a searchlight that bathed the top with light for experiments at West 10,000. Hornig made one last check and left. He was the last man to see Fat Man before Zero Hour. Bush and his assistant prowled the immediate area to be sure that only the authorized arming party was present.

Zero Hour was originally set for two o'clock Monday morning. Hubbard's weather forecast predicted more showers and the shot time was delayed until four. That, too, was moved ahead because of bad weather. At South Shelter, Groves and Oppenheimer consulted on and off until one o'clock. Oppenheimer became so nervous that Groves finally left his tent, asking him to try to get some sleep. Groves himself was tired. He had just spent several hours calming Oppenheimer and trying to force decisions on the test. He was particularly peeved about what he considered to be bad advice given to Oppenheimer by other scientists. Groves joined Conant in a small tent near the South Shelter for some sleep, but the wind picked up and rattled the tent flaps, and neither man slept well.

At four o'clock the rain stopped. Hubbard and Bainbridge waited another 45 minutes for a final report: "Winds aloft very light, variable to 40,000 surface calm. Inversion about 17,000 feet. Conditions holding for next two hours. Sky now broken, becoming scattered."[4] Hubbard seemed pleased with the report; even the winds favored a test. Bainbridge and Hubbard consulted with Oppenheimer and Farrell through John Williams. It was agreed: the test would go on at 5:30 A.M.

Bainbridge and the arming party moved into action. With Kistiakowsky and McKibben, Bainbridge drove to a checkpoint 900 yards south of Zero. There McKibben threw the main bank of timing and sequence switches, with Bainbridge calling out the checklist of tasks under the glare of a flashlight. From there they quickly drove back to the tower, where Bainbridge threw the special arming switch and connected the arming, power, firing, and informer leads. Every step was called by FM radio to Williams at South Shelter: if anything went wrong, or a step was missed, Williams would have a second checklist to consult. A bank of searchlights was turned on. It bathed the tower in harsh lights. The illumination would provide the B-29 reconnaissance planes with a viewing target from the air. On the way back from the tower, Bainbridge broadcast the weather report to outlying shelters and to the teams on Compañia Hill. At South 10,000, Bainbridge unlocked the primary firing switches at five o'clock.

McKibben started the automatic timers at 5:09. After 20 minutes another, more precise timer would automatically take over, and just seconds before the actual connection it would alert another bank of switches to start special instruments across the desert floor.

An urgent call from Kirtland Air Base in Albuquerque brought word from Parsons that the weather in Albuquerque was so bad that the Base Commander would not let him take off. After considerable argument, the planes were released, but they became disoriented by the bad overcast. They arrived at Trinity too late to drop special measurement gauges and, instead, remained observers from a distance.

Everyone moved into his place. The time had come. At South 10,000, Oppenheimer, Groves, Farrell, Bainbridge, Kistiakowsky, and Bush gathered. Hornig sat at the master control panel, and Sam Allison was at the countdown panel. Robert Wilson and his team were at North 10,000; John Manley, Julian Mack, and others waited at the West Shelter. At Base Camp, Fermi, Rabi, Holloway, and Lieutenant Bush took their places on the ground in a shallow depression. A few men around them fussed with their equipment. Bob Krohn made light conversation with Fermi and asked him whether the goggles and welder's glasses were really necessary. Fermi said he didn't know, but that what was going to happen was too important to miss. On Compañia Hill, Teller reminded everyone about the dangers of sunburn and produced an Army Issue ointment. Both scientist and luminary huddled on the ground. Richard Feynman, Klaus Fuchs, James Chadwick, and Geoffrey Taylor made themselves as comfortable as possible on the chilly mountain.

Over Allison's countdown the project's radio channel suddenly blurted the "Star Spangled Banner." Inexplicably, a California radio station had crossed frequencies with Trinity. Forty miles away in Carrizozo, Al and Elizabeth Graves waited in a dingy motel room with a barograph to record shock waves. In nearby San Antonio, Army MP's woke the local restaurateur, Jose Miera, with the inducement that if he came outside he would see something never seen before. Campers from Los Alamos waited on nearby Chupadero Peak, and a few Laboratory wives moved into the mountains above Los Alamos for a chance of seeing the flash of light from Fat Man. At Potsdam, Truman was preparing to see war-ravaged Berlin and Hitler's bunker.

Over Trinity's radio, Allison's calm voice gave the countdown at five-minute intervals. Oppenheimer grew so shaken that he was forced to cling to a post to steady himself. At minus 45 seconds the precision timer took over, and then only Hornig could stop the process. Williams, Bainbridge, and Kistiakowsky rushed outside the shelter. It was all silent except for Allison, who counted down to Zero and shouted, "Now!"

At first, from the darkness, came the light: a stupendous burst of fierce light many times more brilliant than the sun. Instantly the surrounding desert and mountains were bathed in white brilliance. Even those with their eyes closed were able to sense the explosion of light and feel the warmth of it on their bodies. Almost everyone was momentarily blinded and dazed. Those recovering first turned to see through their welder's glasses a huge ball of fire, like the sun, rising from the desert floor in a swirling inferno of reds and oranges and yellows.

The fireball rose from a darker stem and diffused into a less intense light that still lit the mountains behind. The ball of fire became increasingly dull and took on a bluish haze at its perimeter. Everyone stood mutely looking at the play of lights. Almost a minute passed before the shock wave hit and knocked down two men at South 10,000. Shortly thereafter, there was a loud crack, followed by a mighty roar that thundered across the desert, echoing against the distant mountains. Even five minutes later there was still a distant rumbling in the valley.

Everyone was shocked, dazed, awed, and somehow both pleased and terrified. Something fundamental and primordial struck at everyone present. General Farrell, who was with Oppenheimer at South 10,000, thought the results were unprecedented and terrifying. Nothing he or the other men had seen before could compare with this. The light alone lit every peak and crevasse with a beauty and clarity that no one could describe.[5]

Fermi's first impression was that of an intense flash of light followed by a sensation of heat on his body. When he rose to look at Zero he saw the mushroom rise quickly to over 30,000 feet. Very calmly, as he waited for the shock wave, he tore a sheet of paper into small pieces and dropped them one by one as the air blast hit him. By watching the distance they traveled, he was able to calculate roughly the force of the explosion.

Fermi's experiment would have looked like madness everywhere but at Trinity, where people were too awestruck to notice such a small thing. Even on Compañia Hill, dazed observers felt the heat on their arms and faces, and the light seemed brighter than the noonday sun. From Base Camp, Victor Weiskopf saw the smoke ball surrounded by a blue glow; almost certainly he thought it was caused by gamma rays from the radiation. He wondered how much radiation was being emitted. A billion curies? Perhaps more like a thousand billion curies? Joe McKibben, at the control panel inside South 10,000, saw the entire room lit up brilliantly by light coming in from behind the shelter. He ducked quickly outside and remembered that the shock wave had not yet hit. He had the impression that "this thing had really gone big."

Bainbridge also was dazed and awed; it was not like anything he had witnessed before. It was impressive and satisfying and also a relief. He was relieved that he did not have to worry about what went wrong. On Compañia Hill the hushed silence gave way to light applause and finally, after many long minutes, to great relief and loud congratulations.

For some there was hesitation. In that grand and mighty moment there was also a sense of cataclysm and foreboding. For Oppenheimer, it recalled the words of the sacred Hindu book, the *Bhagavad-Gita*:

> *I am become death,*
> *The shatterer of worlds.*

Bainbridge turned to Oppenheimer as the light faded and said that they were now all sons of bitches. William Lawrence, the only journalist present, watched from Compañia Hill. He saw it as the "grand finale of a mighty symphony of the elements, fascinating and terrifying, uplifting and crushing, ominous, devastating, full of great promise and great foreboding."[6]

The light continued to dim, and morning rose gently behind the dissipating cloud. The mighty roar softened to a murmur in the mountains. The sense of relief and success spread. Men in Base Camp, in the shelters and observation posts, all breathed more easily. Suddenly, everyone could release the tension of many weeks in a feeling of great success. Kistiakowsky rushed up to Oppenheimer and put his arm on a shoulder and reminded Oppenheimer of the bet they had made: $10 of Oppenheimer's money to Kistiakowsky's salary for a month if the bomb didn't work. Other men broke out bottles of beer and bourbon.

The explosion had been seen elsewhere. The first flash of light was seen in Albuquerque, Santa Fe, Silver City, Gallup, and El Paso. Windows had been broken in nearby buildings and had been rattled in Silver City and Gallup. A rancher sleeping near Alamogordo was awakened suddenly with what seemed like a plane crashing in his yard. Wives from Los Alamos on Sawyers Hill saw a great flash of light that lit up the trees and produced a long, low rumble. A Forest Ranger in Silver City reported an earthquake to the Associated Press, and other sightings drifted into newspapers in Texas and New Mexico reporting meteors and crashing B-29 bombers. Mary Lapaca of Socorro reported a rocket bomb over her house.

The Associated Press office in Albuquerque soon had a number of queries and reports on a strange explosion in southern New Mexico. Groves had already taken precautions: he had stationed Phil Belcher, an Army Intelligence officer, there to prevent alarming stories from going out. By late morning, the AP man couldn't hold back some kind of story. If the Army had no official story, AP would issue its own. Groves had been prepared for this. Weeks earlier he had ordered the preparation of a news release to be issued from the Alamogordo Bombing Range. The Commanding Officer, Colonel William Eareckson, had been told by his superiors to issue the story upon request from Lieutenant Belcher if it became necessary. Before noon, the AP had the following story.

Alamogordo, July 16–The Commanding Officer of the Alamogordo Army Air Base made the following statement today: "Several inquiries have been received concerning a heavy explosion which occurred on the Alamogordo Base reservation this morning.

"A remotely located ammunition magazine containing a considerable amount of high explosives and pyrotechnics exploded.

"There was no loss of life or injury to anyone, and the property damage outside of the explosives magazine itself was negligible.

"Weather conditions affecting the content of gas shells exploded by the blast may make it desirable for the Army to evacuate temporarily a few civilians from their homes."

New Mexico newspapers ran the story in different versions, and the story appeared in a number of radio shows. No further word was issued by the Alamogordo Base.

The concern for evacuation was real, and it almost became necessary at Trinity. At North Shelter, Dr. Henry Barnett's monitoring crew suddenly found their radioactivity counters clicking wildly. Barnett gave the order to evacuate, and soon trucks and jeeps roared down the road to Base Camp. Later it was found to be a false alarm. Film badges worn by the men showed that no radioactivity had reached the shelter. At another shelter, a radioactive cloud passed over and the men had to put on masks. That, too, passed and by 9:30 A.M. Bainbridge gave the order for men in all observation sites to return to camp.

Groves had hoped to use the period after the test for meetings with Oppenheimer and his men to discuss future plans. As it turned out, that plan was useless. No one had the frame of mind or energy to contend with discussions of the future. Groves remained at Trinity until all danger of radioactive fallout passed. By late afternoon, Bainbridge concluded that there was no further danger from radioactivity and gave the word that all was clear. He returned from South 10,000 to Base Camp a little after three o'clock for food and sleep. Groves joined Bush and Conant and drove to Albuquerque for the trip back to Washington.

Oppenheimer got into a car with a renewed energy that amazed those who saw him.

Shortly after seven in the morning, Jean O'Leary, Groves' secretary in Washington, received a coded message from the General: the test had been a success. She in turn sent a previously prepared message to Secretary Stimson at Potsdam:

Operated on this morning. Diagnosis not yet complete but results seem satisfactory and already exceed expectations. Local press release necessary as interest extends a great distance. Dr. Groves pleased. He returns tomorrow. I will keep you posted.[7]

Although it would be weeks before a thorough assessment of the Fat Man could be made, preliminary studies showed that the explosive results had been greater than expected. Groves could therefore signal Stimson and Truman the next day with even better news:

Doctor has just returned most enthusiastic and confident that the little boy is as husky as his big brother. The light in his eyes discernible from here to Highhold and I could hear his screams from here to my farm.[8]

Stimson was elated. Groves was expecting the Little Boy uranium gun to be every bit as powerful as the Fat Man. The Trinity explosion had been seen even at Los Alamos. What Stimson did not know yet was that Fat Man had exceeded the original prediction of 10,000 tons of TNT; instead, the bomb had produced an explosive force well over 17,000 tons. Stimson now knew, however, that America had the power to crush Japan and barter with Stalin.

At Trinity and Compañia Hill, men began to return to Los Alamos, tired but jubilant. Several hours after the explosion, Fermi was able to enter Ground Zero in a lead-lined tank to capture samples of the radioactive earth. The scene was startling: a large crater 1,200 feet in diameter had been formed; in it, all vegetation had vanished. In the very center was a shallow bowl nearly 130 feet wide and some 6 feet deep. All of the desert sand had been pulverized and fuzed into a dull green glass. Even five miles away scientists could see Ground Zero shimmering in the sun. The steel tower had completely vaporized, and half a mile away Jumbo had been torn from its sturdy metal tower and thrown to the ground.

A week later Oppenheimer met with Groves and Tolman in Chicago to review the Trinity test. Groves and Tolman learned that at ten miles the brilliance of the explosion was that of 1,000 suns, and that enough gamma radiation had been emitted to be lethal to everything within a radius of two thirds of a mile. There was also relief and amusement at Fermi's earlier prediction that with Fat Man there was one chance in thirty of destroying New Mexico and one chance in a thousand of destroying the world.[9]

President Truman waited eight days before mentioning news of the success at Trinity to Stalin. Churchill had been given news of the test on July 16; both the Americans and the British had shared General Groves' lengthy report as soon as it arrived. In response to Truman's casual mention of the bomb, Stalin merely said, "Good," and expressed the hope that it would be used against Japan. Neither the Americans nor the British knew that Klaus Fuchs and David Greenglass had for two years been passing secret information to Soviet agents.

At Los Alamos, the task was now one of delivering bombs. The time had come for Japan.

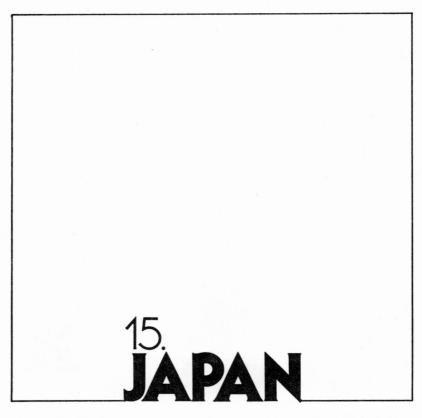

15. JAPAN

The spectacular success at Trinity briefly raised the question of whether the new weapon needed to be used against Japan. To some in Los Alamos, it was as if the real work had been done. The question of use and control, however, had long since passed into other hands. Los Alamos had been created as a scientific *and* military venture; after Trinity, the military's ascending control became complete. Project Alberta, the last big push in Los Alamos before the end of the war, was, after all, a military operation led by a military man.

Navy Captain Parsons' Ordnance Division was as old as the Laboratory. Project Alberta had its genesis in 1943, when plans were laid for the eventual delivery of combat weapons into the hands of the military. Ordnance, more than any other division in the Laboratory, could never forget its ultimate task. Other divisions, and other men within those divisions, could afford to come to Los Alamos and luxuriate in Oppenheimer's heady atmosphere of big science. Parsons and his men could not.

Some of their earliest work focused on designing and testing bomb casings. With so much of the Laboratory's thinking still so fluid during 1943, the first few shells designed by Ordnance were preliminary at best. As Los Alamos work narrowed to Little Boy and Fat Man weapons, Parsons was able to refine their external casings and reasonably perfect their characteristics in flight. Groves had already selected the B-29 as the aircraft for Los Alamos to use. Fifteen B-29s were built at the Martin Aircraft Plant in Nebraska according to preliminary Laboratory requirements that called for dropping one long or one very fat bomb. Parsons had assigned two men to work with Martin to incorporate projected bomb requirements.

The first lot of B-29's was available in October 1944 and sent to a newly created Atomic Bomb Group stationed at Wendover Army Air Base in Utah. For security purposes the base was referred to as Kingman, or more simply, W-47. Groves arranged for a special bomb group to be created and stationed there. Colonel William Tibbits was made commanding officer of the new 509th Composite Group.

The first tests at Wendover began in October with dummy elliptical shells and long tubes made to represent the Fat Man and Little Boy casings. The first B-29's received and put into service were very unsatisfactory. Groves and Parsons ordered a new lot of fifteen planes. These arrived in the spring and had new innovations: fuel injection and electrically controlled propellers. There were also more modifications in the bomb bay for carrying the new weapons, and all the armor was stripped from the plane except for the tail turret. Bombing accuracy improved tremendously with the new airplanes.

With its creation in March 1945, the new Project Alberta incorporated several small teams at Los Alamos, including the Delivery Group. Very quickly it began to draw on other divisions for men and assistance. Norman Ramsey and Norris Bradbury assumed deputy responsibilities for scientific and technical activities. Three new units were created within Alberta. The first was a Headquarters Staff, located at Los Alamos and directed by Colonel Ashworth. Headquarters had overall responsibility for coordination. The second unit was the Weapons Committee chaired by Ramsey; and the third unit was a miscellaneous collection of individual staff members or small teams whose work was related to weapon delivery.

Alberta's work became increasingly more demanding with each new success in Los Alamos in 1945. With the weapons freeze in February, Parsons could complete planning and preparations for overseas. Alberta had already stopped planning for Europe, and only Japan remained. Colonel Ashworth visited Tinian Island in February to conduct a site survey. After his return, plans for an isolated compound were drawn up for one end of Tinian. Navy Seabees were consulted and were assigned construction duties.

At Tinian, known as Destination, the Alberta team would occupy part of the area near the ocean that belonged to the First Squadron of the special 509th Group. Four quonset huts with air conditioning were built as laboratory buildings for use with delicate instruments. There were also storage buildings, a shop, and one administrative building. A mile away were three widely spaced assembly buildings in which Fat Man and Little Boy would be put together for the last time. The four quonset huts were enclosed within a special high-security fence and were heavily guarded.

In June 1944, the need for overseas personnel was raised for the first time. Oppenheimer thought it would be better to solicit volunteers for the work because of the location and danger. He told his division leaders on June 13 that it was apparent that Los Alamos would eventually be engaged in overseas work. Men would be needed for assembly, fuzing, loading, and checking circuits and test equipment. Prospective volunteers needed to be warned of the danger, the travel to "advanced locations," and the likelihood that it might mean transfer to military status. Oppenheimer urged his staff to ask for volunteers and not to apply pressure. It was clear, he said, that "adaptability, high morale, ability to do work under unusual and difficult conditions, stamina, tact, and willingness to accept necessary military discipline should all be given weight in the selection."[1]

Oppenheimer later reported to Groves that the response was "by no means unanimously favorable, but a considerable number of men have signified their willingness to go."[2] Oppenheimer did not think the Los Alamos field force should be entirely civilian, but that those who went should go under the University of California's contract. He recommended a largely civilian force, under contract to the government but given suitable military rank and insurance, and with provision made for their families during the period of overseas duty. A preliminary list of volunteers included Kistiakowsky, Donald Hornig, Dr. Hemplemann, Alvarez, Ramsey, and Harold Agnew.[3]

Recruiting was delayed through December, although it was expected that the ship and air crews would leave May 1, 1945. Crew appointments were made in April and May, and included Parsons as Officer-in-Charge and Ramsey as Scientific and Technical Deputy.* Observers and special consultants included Alvarez, Serber, Penney, and James Nolan. Team or group members included Harold Agnew,

---

*Colonel Ashworth was Operations Officer; Warner headed the Fat Man assembly team; Birch headed the Little Boy team; Stevenson the Electrical Detonator team; Morrison and Baker the Pit Assembly group.

Raemer Schreiber, and thirty-four others, mostly enlisted men. General Farrell went as personal representative of Groves.

This selection of civilian and military men became part of the First Technical Service Detachment, and it departed to Army housing and various services on Tinian. While they all remained on the Los Alamos payroll, they were provided *per diem* pay and uniform allowances, and given additional insurance policies. For the moment at least, everyone was given a rank and a uniform with the rank purportedly consistent with civilian salaries. None of them, however, was permitted to carry a gun.

The first shipment of supplies and bomb components arrived in May. Not until July were the uranium and plutonium cores sent overseas under special arrangements code-named BRONX. Oppenheimer and Parsons had planned for both cores to be sent separately. The plutonium was to be sent by airplane and the uranium by naval ship. The plutonium package consisted of the active core, the high-explosive lenses, the electrical firing unit, and the detonators. It was agreed that the plutonium would not leave the continental United States until after the Trinity test.

Both packages arrived safely in Tinian. The U 235 had been carried by the U.S.S. *Indianapolis*. Unknown to the men on Tinian, the *Indianapolis* was torpedoed by a Japanese submarine on its return trip. It was a slow death: the ship sank slowly, and most of the men dived into the ocean to await rescue. Help did not come for three days, and hundreds of men drowned or were eaten by sharks.

Unaware of the tragedy, the men at Tinian used the remainder of July and the first few days of August to run tests and to examine and reexamine each component of the bombs. Test flights were made by Tibbits' men, with heavy weights to simulate the bombs themselves. For the most part, the scientists engaged in busy work or wandered around the island.

A few men from Los Alamos brought along American-made goods to use as barter or exchange on Tinian. Luis Alvarez, for example, brought along a case of American bourbon; Harold Agnew brought a case of soap. At the time, Pacific conch shells were particular favorites of American GI's. To Alvarez's dismay, American liquor was cheap and available everywhere. Soap, however, was greatly in demand, and Agnew was able to acquire a number of the highly prized shells.

The first days of August brought an increasing anxiety to the Los Alamos and 509th teams. Parsons and Ramsey began to cable General Groves for permission to drop. They had reason to be concerned: bad weather had been predicted for Japan in early or mid-August. The 509th was ready now.

Trinity had been very useful in enlarging the thinking of Los Alamos for the use and delivery of the bomb. Parsons received a memorandum from Bradbury and Kistiakowsky the day after Trinity with suggestions. Both men told Parsons that the immediate flash of light was more than expected. This discovery suggested that if enemy troops could be drawn into visual range of the bomb considerably more damage might be accomplished. It was their feeling that "no one within a radius of five miles could look directly at the gadget and retain his eyesight."[4] It was not known whether the effects would be temporary, but some device such as a siren or flashing light could be dropped from a plane in advance of the bomb itself. Such a thing had happened accidentally at Trinity. One man at Base Camp had been asleep in the open and was awakened suddenly by the first flash of light. Instinctively he looked in that direction and was blinded for several hours. He had missed the most intense light, but nevertheless the effect had stunned him. The memo concluded with a wry note that the trick might not work indefinitely.

A few days later, Oppenheimer reported further implications to Groves and Parsons. The Fat Man or Little Boy could be expected to produce a blast between 8,000 and 15,000 tons of TNT. The fuzes were set to detonate at 1,800 feet above the ground, and radioactive contamination was not expected to reach the ground. This suggested that care should be taken by air crews not to look in the direction of the blast for some time. The ball of fire would persist longer than at Trinity, because there would be little dust involved. Oppenheimer gave the Little Boy a good chance of "optimal performance"; only a 12 percent chance of less than this; a 6 percent chance of an explosion under 5,000 tons; and a 2 percent chance of one under 1,000 tons of TNT.[5]

There was also the question of where in Japan to drop the bomb. To Groves, the issue was primarily a military or strategic one, but he nonetheless solicited the recommendations of Los Alamos, and there was an exchange of information on possible sites. The Laboratory was removed from most of the thinking on strategy in Washington but met several times in May with military officials to discuss potential Japanese targets.

The Target Committee, as it was called, first met in Washington on May 2. Brigadier General Lauris Norstad presided at the meeting with Groves and Farrell attending from the central Manhattan

District office. Oppenheimer could not attend, but sent John von Neumann, Robert Wilson, William Penney, and Joyce Stearns from Los Alamos. It was clear to everyone that visual bombing was a necessity; existing radar equipment was not sophisticated enough to pinpoint targets. Weather over Japan was the obvious next question. The best available information suggested that July was less than favorable for precision bombing over Japan. August could provide perhaps six or seven days of clear weather, and September was generally unfavorable. There was clear pressure to act in early August.

Hiroshima appeared the first choice as a target city, since it was virtually untouched on the 21st Bomber Command's priority target list. The 20th Air Force, which included the 21st Bomber Command, had general jurisdiction over Japan. For some time, American bombers had systematically sought to bomb major cities in Japan, and other targets for the atomic bomb therefore included Yawata, Yokohama, Tokyo, Nagoya, Osaka, Kobe, and Nagasaki. Penney was given the task of calculating the prospective bomb damage in each city, and Stearns was asked to compile further target data from the Joint Army-Navy Target Group. A second meeting was planned for Oppenheimer's office in Los Alamos on May 10 and 11.[6]

The two-day meeting reviewed the Washington discussions and took up technical questions and target selection. Optimal heights for detonation of Little Boy and Fat Man were presented, as well as recommendations for jettisoning the bombs should it become impossible to make a clear drop on a target city. The weapons being prepared for Japan comprised the entire American arsenal of atomic bombs. In case a bomber had to return to Tinian with its payload, the first priority was to land with the bomb intact and with the greatest caution. If a "normal" return landing could not be made, it would be necessary to jettison the Fat Man into shallow water from a low altitude. Little Boy, however, could not be dropped into the ocean, since water leaking into the nuclear core would set off a chain reaction. Such a reaction could be explosive, and would severely damage Tinian and nearby American installations. For Little Boy, the best emergency procedure would be to remove the explosive powder from the gun and crash land the airplane. The committee members agreed that some sort of "instruction book" might be helpful to the pilot in case of an emergency landing.

At the end of the second day, Stearns made his report on the Laboratory's target recommendations. His report mirrored what the military was considering: important targets in large urban centers, capable of being heavily damaged, and likely to be unattacked by conventional American planes by August. The Army Air Force had provided

Stearns with a list of five targets in Japan which they would be willing to "reserve" for the Laboratory's special use.

The first was Kyoto, which was a large industrial center with over a million residents. It was thought that Kyoto was particularly attractive because it was an intellectual center for Japan and presumably the residents would be better able to appreciate the significance of an atomic weapon. The second target city was Hiroshima, an important army depot and port of embarkation in the middle of an urban industrialized area. The large hills surrounding the city were thought to be effective reflectors for the blast and shock waves.

The third city was Yokohama, which was attractive because as of May it had not yet been bombed by the Air Force. Aircraft and electrical equipment factories and oil refineries were all in the area, and damaged industries in Tokyo had already begun to move to Yokohama. Unfortunately, it was one of the most heavily defended cities in Japan.

The Kokura Arsenal was a fourth possibility because it was one of the largest arsenals in Japan. The center contained a large amount of light ordnance, antiaircraft, and beachhead defense materials. The fifth target was Niigata, which was an important port of embarkation on the northwest coast of Honshu. For a while the Laboratory considered the Emperor's Palace in Tokyo. It was finally rejected when it was agreed that the military might have some reason to avoid this area in terms of future relations with Japan. Kyoto and Hiroshima both received AA ratings* from the Los Alamos committee, while Yokohama and Kokura received only an A rating; Niigata was given a B.

Stearns agreed to collect more data on each target, as well as to brief the Air Force liaison at the Laboratory. The committee felt that it was most important to use the weapon to gain psychological as well as military advantage. The use of the weapon needed both Japanese and international attention drawn to it to maximize its terror and destructiveness; this would encourage, or force, capitulation and serve as warning to the rest of the world. Kyoto was particularly valuable as an intellectual center, Hiroshima had the greatest potential for physical damage, and the Emperor's Palace had the greatest significance to the Japanese people.

The committee also discussed the possibility of following the bomb attack with a heavy incendiary mission. This concept had the advantage of catching the enemy when their recovery from the bomb would have left them severely debilitated and with fire-fighting ability paralyzed. So little was known, however, about the phenomena as-

---

*"AA" identified the best choices, "A" second choice, and so on.

sociated with the weapon—fireball, radioactive cloud, and weather—that it was decided that an incendiary mission should take place no earlier than the day after the nuclear attack. The meeting concluded with Parsons and Ramsey agreeing to meet with Groves to pursue the matter.

Groves and Farrell had undertaken their own planning and along with representatives from the Air Force, had agreed on three cities as first choices: Kyoto, Hiroshima and Niigata. Industrial targets would not be sought, but rather the center of each city would be ground zero. Stimson was not pleased with Kyoto as a target. It was an ancient city and of great cultural value to the Japanese people. He refused to allow Kyoto to be a target. As a result, Groves later submitted a revised list of targets to General Marshall that included Kokura as a replacement for Kyoto.

These target recommendations assumed that the atomic bomb would in fact be used. Stimson's Interim Committee on S-1* became a sounding board for discussions on use of the bomb. In three years, the bomb's use was never discussed as anything but fact. The committee acted on the assumptions that work on such a bomb by Germany necessitated the American version; that it could end the war with Japan; and that more than two billion dollars had been spent to achieve the bomb and its implications for the postwar world.

Stimson still sought to incorporate nonmilitary views into his own thinking. He called a meeting of the Interim Committee for May 31 with a carefully structured agenda that excluded any call for discussion on *whether* the bomb should be used. In his meticulous wartime diaries he noted that the meeting sought to reinforce that "we were looking at this [the use of the bomb] like statesmen and not like merely soldiers anxious to win the war at any cost."[7] Ostensibly the agenda called for a discussion of topics such as temporary control of the bomb, postwar research, controls, and future nonmilitary uses.

The meeting began with the Committee present, as well as with Arthur Page and Generals Marshall, Groves, and Harvey Bundy as guests. Stimson began by suggesting that the bomb had implications beyond the war against Japan. Arthur Compton and Oppenheimer then outlined the next several generations of weapons. Oppenheimer stressed that development of these weapons was merely a question of

---

*The Interim Committee had been created by Secretary of War Stimson in 1943 as an advisory body on the Manhattan District. Stimson acted as its chairman. It had as members Ralph Bard, an undersecretary in the Navy Department; William Clayton from the State Department; Bush, Conant, Compton, Oppenheimer, Ernest Lawrence, Arthur Compton, and Fermi.

time. Beyond the Trinity bomb, for example, was a second generation of bombs with explosive forces between 50 and 100 kilotons of TNT. Utilizing these weapons as ignitors, it would be possible to construct weapons with explosive forces between 10 and 100 million tons of TNT.[8]

It was clear to Stimson and the others that the Manhattan District as an entity would have to remain beyond the war, and that research and development must continue and must provide for a stockpile of weapons. The subject of Russia was also raised. Oppenheimer repeated Niels Bohr's belief that Russia must be told about the atomic bomb without details. There was some agreement over this approach until Secretary Byrnes, who had joined the meeting, objected strenuously. After some lively discussion, there was a consensus that the United States must retain world leadership in atomic matters.

Stimson thought it wise to break for lunch. It was during lunch, however, that the question of the bomb's actual use—the need for its use—arose among committee members. One possibility discussed was the use of the bomb in a nonmilitary demonstration. Such a demonstration might be successful in moving the Japanese government to capitulate, particularly if the display was dramatic enough. Oppenheimer was forced to suggest that such a display could not surpass the bomb's effects on physical structures such as buildings. Arguments against use in Japan were eliminated one by one. A surprise drop was essential to prevent counteraction by the Japanese. A single explosion, say, on an island, might not convince the Japanese leaders that the explosion was caused by a single weapon.

The meeting formally began again and carried on until the late afternoon. Stimson was required to alter the agenda. It was finally agreed that the bomb be used *without* warning against the Japanese and that it be used against a target that would impress as many of the enemy as possible. The meeting confirmed the belief that the bomb would be used. Moreover, it recommended that both Little Boy and Fat Man be used on different targets. Although the war in Germany was still going on, the use of the bomb there was never raised.* A week later, Stimson passed along the committee's recommendations to President Truman.

Target selection then continued until late July. Oppenheimer reported to Groves that a uranium Little Boy would be available soon after August 1 and that another Fat Man would be

---

*Sherwin, in his A World Destroyed, suggests that racism was not a reason, as has been charged by some historians. Rather, he suggests that the war against Germany was virtually over, that the war against Japan was an "American" war, and that the bomb was an American product. Moreover, use of the bomb in the East was logistically safer than use in Europe.

available after August 5. Oppenheimer also reported that the bombs should produce good results in use. Calculations based on Trinity suggested that these bombs would deliver a force between 8,000 and 15,000 tons of TNT, and that the fireball would be greater in brilliance than the one at Trinity. On July 24, the target list was finally narrowed to include four cities: Hiroshima, Kokura, Niigata, and Nagasaki. This decision was the result of much haggling among Stimson, General Henry Arnold, and others in Washington. Stimson had at least been relieved to learn that Truman also favored sparing Kyoto because of its historical and cultural value.[9] Little Boy and Fat Man were readied for Japan.

At Destination, Parsons and Ramsey finally received permission from Washington to drop the bombs as soon after August 3 as the weather permitted. Parsons was ready, but was forced to wait three days until the weather cleared. Parsons had perfected a timetable much like Bainbridge's Trinity schedule: unpack Little Boy and assemble for tests and inspection by July 29; assemble with fuzes and informers and by August 4 assemble with explosive lenses without batteries. Twelve hours before leaving, the unit was to be packed with its electrical system complete and "hot."

As they waited for the weather to clear, Parsons and Ramsey decided not to take off with the Little Boy entirely assembled and armed. An accidental crash on takeoff could cause enormous destruction to Tinian if the bomb were to detonate. There was no way to make the bomb entirely safe, but at least a partial safeguard could be taken: the bomb would be loaded on the airplane without is propellant charge. With the plane safely in the air and en route to its target, Parsons could then insert the charge through the opening in the breech block. The bomb bay of the Enola Gay—the B-29 flown by Tibbets and scheduled to carry the Little Boy—was modified with a small platform for Parsons to stand on to make the final assembly in flight. Final arming would occur when Parsons pulled a small wooden plug four inches long from the body of the bomb. Later, if successful, Parsons planned to present the plug to General Groves as a trophy.

On August 5 word came that the weather the next day would be good over Hiroshima, the primary target, as well as over two secondary targets. Kokura had been made a secondary target choice because Army Intelligence suspected that an Allied prisoner-of-war camp was located nearby, although its exact location and number of prisoners were not known. Kokura was a better target than Hiroshima for bomb damage because of its terrain, but there was hesitation to bomb an area containing Allied prisoners of war. The Army believed that the Japanese had purposely moved prisoners into camps in major target cities to discourage Allied bombing. Eventually the War Department was forced to rule that such camps could not be a factor in deciding bombing targets.[10]

The same day word came from General Curtis LeMay to proceed on August 6. The Little Boy was placed in its special cart in Assembly Building 2 and driven to the Enola Gay. The operation was watched carefully by the Los Alamos teams under a heavy contingent of Security Police. The final briefing was held at midnight and the crew was photographed in front of its aircraft. Paul Tibbets was commander; Major Thomas Ferebee, bombardier; and Captain Ted van Kirk, navigator. Parsons, although a Commander in the Navy, represented Los Alamos and was a bomb commander.*

At 0245 the Enola Gay took off from Tinian with an escort of fighter and reconnaissance planes. It took almost the entire runway for the heavily laden aircraft to lift off. Thirty minutes later, Parsons inserted the explosives charges and completed the final assembly of Little Boy. At 0605 the Enola Gay flew over Iwo Jima and headed toward Japan. An hour and a half later, Parsons inserted the red plugs that armed the bomb to detonate after being released from the airplane. The plane leveled off and Parsons once again checked the fuzes. Shortly after 0900 Hiroshima was in sight.

At 9:14:30 August 6, Ferebee dropped the bomb. The sudden release of the bomb caused the plane to lift up a dozen feet. A few minutes later there was a brilliant flash, similar to the one at Trinity, that lit up the interior of the Enola Gay. Seconds later the aircraft was severely jolted twice by shock waves. The large ball of fire seemed to churn within itself and quickly changed to swirling purple clouds and boiling flames that surged upward. The mushrooming cloud reached a height of 30,000 feet in less than three minutes. It rose another 10,000 feet and flattened itself on top. A combat plane 360 miles away was able to see the cloud at an altitude of 25,000 feet. From an observer plane called the Grand Artiste, Harold Agnew took the only movies of the Hiroshima bombing with a 16-millimeter home movie camera.

The effects of Hiroshima were awesome. A city of 300,000 was devastated. The very center of the city had been leveled and more

---

*The crew also included Capt. Bob Lewis, Lt. Jacob Beser, Lt. Morris R. Jeppson, assistant to Parsons, Sgt. W. E. Dusenbury, Sgt. Robert R. Shumard, Sgt. Joe A. Stiborik, PFC. R. H. Nelson, and Sgt. G. R. Caron.

than 60,000 buildings had been destroyed. Shortly after the bombing, great waves of fire had swept across the city. While a few modern concrete buildings survived as shells at the outer limits of the city, the force of the explosion had been so great that it tore loose handrails on nearby bridges.

The day after the Hiroshima strike, General Farrell received instructions from the War Department to begin a propaganda campaign against the Japanese. Following a previously prepared scenario, Farrell and his assistant, Major Moynahan, ordered the dropping of six million leaflets on heavily populated Japanese cities. The leaflets described the new and powerful Allied weapon that had destroyed Hiroshima. They were supplemented by Japanese-language newspapers with photographs and details of the Hiroshima explosion. Over Radio Saipan, the Army began brief broadcasts every fifteen minutes describing the bombing. The campaign continued until the Japanese government began negotiations.[11]

Three days later, on August 9, another strike plane left Tinian with the world's second Fat Man on board. The bomb had been quickly assembled and checked: explosives assembly, plutonium core and initiator, and the X, or detonator, assembly. It rode snugly within its egg-shaped metal shell.

Fat Man was of such a shape as to make loading more difficult than that of Little Boy. With its stabilizing fins, the bomb was 12 feet long and 62 inches wide and weighed nearly 10,000 pounds. Fat Man had to be rolled to its airplane on a special carrier. After it was under the forward bay of the B-29, special horizontal jacks would adjust the cradle to accommmodate the large bomb. The entire process took almost 25 minutes. Unlike Little Boy, there was nothing that could be done to make the Fat Man safe before liftoff; the bomb and its crew would have to take off and hope for the best.

The silver B-29 called Bock's Car was commanded by Major Charles Sweeney. Sweeney's own plane, the Grand Artiste, was fitted with special photographic equipment and was unable to accommodate the heavy Fat Man. Sweeney discovered at the last minute that a fuel tank with 600 gallons of gas was unusable. He decided to risk it anyway.[12] Assisting him were Captain Kermit Beahan as bombardier and Colonel Frederick Ashworth taking Parsons' place as bomb commander.*

Bock's Car took off at 0347 and shortly thereafter Ashworth began the arming procedures. He connected the red plugs to the deto-

---

*The crew also included Capt. James van Pelt, Lt. Charles Albury, Lt. Jacob Beser, Lt. Philip Barnes, assistant to Ashworth, Lt. F. J. Olivi, Sgt. John D. Kuharek, Sgt. Edward Buckley, Sgt. Albert T. DeHart, Sgt. Raymond Gallaher, and Sgt. Abe Spitzer.

nators and checked the condensors for leakage. The plane and its escort rendezvoused over Yakashima. The weather report indicated that Kokura had good weather with low clouds. The secondary target, Nagasaki, also had good weather, but promised increasing cloudiness. Kokura was chosen, but as the plane and her escorts arrived they found the city covered by heavy fog and smoke. Several runs were made, but Beahan found it impossible to locate the aiming point. After 45 minutes over Kokura, the aircraft was forced to proceed, with its fuel low, to its second target, Nagasaki. At 1130, Bock's Car made contact with Nagasaki by radar. Twenty minutes later, the bombardier located the target in a visual bombing run and released the Fat Man. It was 11:50:20.

There was the same swift death for Nagasaki. A great blast of light and shock waves hit Bock's Car. The aircraft slowly circled the dying city as the ball of fire grew and changed colors in a great rising cloud. At five minutes after twelve o'clock, the B-29 left the city for Okinawa to land and refuel. Photographs the next day showed that only 44 percent of the city had been destroyed, although the effects were more spectacular than those at Hiroshima. The blast had been of greater power, and many reinforced concrete buildings and metal structures had been demolished. The larger distance of destruction suggested that the energy blast had been nearly twice that of Hiroshima. The blast had even followed the steep hills and ravines that surrounded the city. The exact point of the explosive burst could be determined by the effects of charring on remaining telephone poles. Unknown to the Americans, the Japanese had placed a prisoner-of-war camp near Nagasaki and a few men—two of them American—became the only Allied victims of the atomic bomb.[13]

It was the end of Japan's war.

Los Alamos was jubilant. The day of Hiroshima Oppenheimer called the Laboratory together in the auditorium in Delta Building. On the podium he clasped his hands together above his head in victory. He read from a message flashed by Parsons 15 minutes after the drop:

Clear cut results, exceeding TR test in visible effects, and in all respects successful. Normal conditions continued in aircraft after delivery was accomplished.

Indeed, Hiroshima was successful: 60 percent of the city was flattened, 37,000 persons were missing, and 78,000 were dead. Three days later,

Oppenheimer passed the word on Nagasaki: less physical destruction, but more than 100,000 persons killed and injured. The city had burned in a raging storm for more than a day.

Los Alamos prepared a third Fat Man for shipment to Tinian. On August 10, Groves wrote General Marshall that this weapon would be ready for use on a target in Japan after August 24. Marshall returned the letter the same day with a handwritten note penciled on the bottom:

It is not to be released over [*Stimson's emphasis*] *Japan without express authority from the President.*

G. C. Marshall[14]

Groves was compelled to cancel the shipment. Nagasaki had been enough to convince Japan that the first bomb was not an accident. With Pacific islands in the hands of the Americans, there was not a city in Japan safe from the new weapon. On August 14, the Japanese surrendered.

When word came, all work at the Laboratory stopped. The Tech Area siren sounded, and everyone near a horn blew it. Kistiakowsky set off a round of old explosives in a canyon below the mesa. Scientists fled from their laboratories into the streets. The good spirit was contagious. All over Los Alamos, as all over America, people moved to one another to celebrate the final victory, and the day wore on into a night of revelry and good cheer.

On the same day as the Hiroshima bombing, Oppenheimer received a telephone call from General Groves in Washington:

G: I'm proud of you and all of your people.

O: It went all right?

G: Apparently it went with a tremendous bang . . . . Yes, it has been a long road and I think one of the wisest things I ever did was when I selected the director of Los Alamos.

O: Well, I have my doubts, General Groves.[15]

For Los Alamos, it was the end of an exciting journey. Their weapons had ended the war. It was an odd feeling: euphoria mixed with sadness, because now both Los Alamos and their lives would change. They were no longer isolated. Hiroshima and Nagasaki had altered that, for Los Alamos was inexorably linked to the strange success in Japan.

16. CHANGE OF GUARD

Success in Japan created more than a new age; it established a new reality for Los Alamos. Suddenly everyone had to confront the prospect of unemployment. Work at the Laboratory was ending. Every day brought the conclusion of one task or another, and everyone had the words "winding down" in his vocabulary. For senior men—Fermi, Allison, Bacher, and others—there could be a return to their academic posts at large universities. For military men like Parsons there was still a service career to continue. For most other men, including those whose scientific careers had begun at Los Alamos, there was only anxiety over the future of the Laboratory and the prospect of jobs.

Oppenheimer's future was equally uncertain. His desire to return to California and teaching was becoming increasingly offset by an awareness of his central role in weapons development. Where he longed for the familiarity of teaching, he also found himself drawn to the great power struggle in Washington over control of nuclear weapons. Oppenheimer, like others, shared the belief that the work of Los Alamos had been useful in concluding the war. He, better than anyone else, realized the mixed nature of the success. He wrote Groves and expressed his feelings on the future of Los Alamos and his desire to leave with the end of the war. Oppenheimer wrote then as the director of an unknown laboratory hidden in New Mexico. In August, with Hiroshima, he was thrust from his relative obscurity into prominence. This sudden emergency, with its social and personal costs, removed him even more from the sweet isolation of academic life.

For most of Los Alamos, however, there was "the future." In Laboratory language, the problem was called reconversion. Los Alamos, after all, had been created as a "temporary" laboratory for war work. The problem of postwar employment promised to be particularly acute for younger men, some with bachelor's degrees and some with just a few years of college or technical training. Most of 1943 and 1944 was spent hard at work and completing work on the bomb, and returning to civilian life seemed distant at best. By the end of 1944, however, the question was raised more often—first among friends, and then with Oppenheimer and his leaders.

At the December 7, 1944, meeting of the Administrative Board, Oppenheimer asked his board members to consider the problem of helping project personnel locate jobs after the war. He felt that the Laboratory must assume this responsibility because Los Alamos could not immediately disengage its work at the war's end, and other projects could. This delay would give a strong advantage in the job market to personnel in other war efforts. Moreover, Oppenheimer pointed out that the severe security at Los Alamos made it very difficult for men to make contacts on their own. It was a fact, he said, that many men at Los Alamos had never worked anywhere else.

Oppenheimer's discussions with the board reflected the growing concerns among Laboratory staff. Already a number of individuals had approached Major de Silva, the Laboratory's security officer, to discuss the process of making applications to other firms without violating security restrictions on outside communication. De Silva suggested that perhaps a central office for applications could be created as an aid to men seeking jobs elsewhere. Tolman, OSRD's representative at Los Alamos, asked that a questionnaire be circulated around the Laboratory to survey the talent. Oppenheimer did not think that this would necessarily be sufficient for the job of placing so many employees. At the very least, he argued, the problem of postwar employment needed to be raised openly with project staff.[1]

Concerns over future employment lessened—at last vocal discussion did—as work in the spring became frantically concerned with preparations for Trinity and Japan. Success in the Jornado created a sense of relief and euphoria that lasted a few weeks. Very quickly, though, with the bombing of Japan at hand, the twin questions of the Laboratory's future and the further employment of each man again became serious. As the Alberta crews at Tinian prepared for the first drops, Oppenheimer's role became increasingly complex. He was faced with concluding the present work at Los Alamos and charting the Laboratory's future into uncertain times. Underneath, like a dark and swift current, was the responsibility for leading both scientists and mankind into a suitable détente with the new weapon.

Oppenheimer tried to assess the situation. In May he wrote a lengthy letter to Groves to establish what he called "cognizance" of the Laboratory's future. Obviously, he wrote, Los Alamos would sustain its efforts to continue the production of the Trinity-style Fat Man bombs, as well as the perfecting of the uranium bomb. The July and August deadlines could still be met. It was important, Oppenheimer urged, to renegotiate contracts with existing firms, as well as establish contracts with new firms, to assume what had become major arms of Laboratory activity. Specifically, Oppenheimer urged that components such as the polonium initiator, the plutonium cores, and explosive lenses be manufactured by commercial industrial firms. He had never thought these items to be part of the Laboratory's "true" duty; only the exigencies of war and heavy pressures from Groves had made them a part of Los Alamos work. Resolution of the Laboratory's future, Oppenheimer urged, could be made more easily by eliminating these activities.

Oppenheimer was clearly trying to separate functions that had evolved over the Laboratory's three-year history. He was careful to indicate, however, those aspects of nuclear research that should be continued. In fact, he urged that some "centralized" government agency be created to continue work in the nuclear field, and that it report to the President of the United States. It seemed clear to Oppenheimer that Los Alamos had served its purpose, and that to continue the Laboratory "in its present form," however, was a mistake. He was to alter his thinking to a large extent within only a few months.

He did feel that some members of the Laboratory should—and could—be persuaded to continue weapons work. "I think," he wrote, "that there will have to be a very great change in the way in which the Laboratory is set up and very probably an actual shift in its physical location." Oppenheimer believed that most of the staff of Los Alamos thought of themselves as occupied with a war job only, and did not plan to continue work on the weapons after the war. "I also know," he added, "that the whole organization, temper, and structure of Site Y Laboraories is singularly unsuited for peacetime perpetuation." His letter ended with a declaration of intent to leave:

In particular, the Director himself would very much like to know when he will be able to escape from these duties for which he is so ill qualified and which he has accepted only in an effort to serve the country during the war.[2]

Oppenheimer spoke more for himself than for his staff. He could see an end to Los Alamos, in delivery of the bombs as much as in the makeshift nature of the Army buildings. If work at Los Alamos could be ended or shifted to a consortium of private business and government agency, perhaps the mesa could be purified and returned to its natural setting.

Oppenheimer's view was at least honest, if not somewhat naive. It did not occur to him, for example, that much of his staff might well want to stay on in Los Alamos. Where could they find work as stimulating and challenging as that of the last three years? What university or industrial laboratory could provide both the intellectual and geographical atmosphere of Pajarito? Ultimately, the struggle for control of the bomb—and therefore of the future of Los Alamos—was being played in Washington. Groves, who fancied his influence in such quarters to be great if not critical, had little doubt in mid-1945 that the new bombs would remain under military control for some indefinite period; Los Alamos would stay as part of that control. Roosevelt, Truman, Secretary Stimson, and others saw the new weapons as part of a greater military and political arsenal. Trinity had given birth to the atomic era, but Hiroshima and Nagasaki had removed the baby from its parents forever. Science and Oppenheimer had lost command. This loss was to be the center of Oppenheimer's life and work for nearly the next decade. However much he sought release from the "burdens" of Los Alamos, or perhaps from responsibility for the bomb, he was caught in a curious and captive bifurcation: between opposing further weapons research and being the spokesman for it, and between a desire to return to the comforts of academic life and the leading public role as father of the atomic bomb.

Oppenheimer was not alone in sensing a new social responsibility. There was an emerging concern among scientists of the Manhattan District for a voice in shaping nuclear policy. For the farsighted few, there was even a concern for postwar international controls. While discussions on the use and import of the bomb occurred in all Manhattan projects, they were most clamorous at the University of Chicago's Metallurgical Laboratory.

Bush and Compton had been approached more than once

by concerned scientists and sought, whenever possible, to reassure them that their concerns were being heard and incorporated into American policy. This was less than true, particularly as the summer of 1945 seemed to promise delivery of the bombs from Los Alamos. Both men were aware that the military's influence was dominating political discussions. This was tellingly clear during the May 31 meeting in Chicago when Fermi, Oppenheimer, Lawrence, and Compton met with Bush, Stimson, and military advisers to weigh alternative uses of the bomb in Japan. Late in the day Bush and Compton asked their fellow scientists to submit research concerns, as well as ideas for postwar controls of atomic weapons. The meeting ended with a request that these ideas be submitted in writing within a few weeks. Compton especially felt the need to have his colleagues share their thinking as a contrast to military perceptions.

Compton went one step further and asked leading scientists at the Metallurgical Laboratory to submit their ideas on use and postwar controls. The request tapped a mounting and ill-concealed anxiety. The reply was full and strong: military use of the bomb against Japan would serious jeopardize, if not prevent, postwar agreement among nations to share control and responsibility for such weapons. Moreover, they urged an end to secrecy; such restrictions, they argued, would similarly hamper exchanges between scientists and policy makers.

For the most part, these discussions between scientists and military and policy leaders were carried on in secrecy and without the inolvement of most leaders of Manhattan laboratories. As work in Chicago wound down, for example, more and more individuals found themselves with time to discuss the course of their work and the implications of the new weapon. These discussions were without benefit of knowledge of Washington politics. A few men, like Niels Bohr and Leo Szilard, had remained active throughout the war in advocating a close collaboration between scientists and military and political officials. Both men had urged work on the new weapon as a safeguard against Nazi developments, but both felt strongly that secrecy should end as soon as possible. As the war neared its end, they urged that the atomic bomb not be used as anything but a threat. Niels Bohr parlayed his international reputation to gain audiences with both Roosevelt and Churchill. Both men were made victims of impassioned but lengthy monologues that only mitigated the listener's interest. Churchill raged out of his meeting with Bohr to ask a startled secretary, "Who was that man?"

Leo Szilard was the individual responsible for persuading Einstein to write Roosevelt initially in 1939 on the prospects for an atomic bomb. During the war, he became increasingly concerned over the fate of the weapon. In July 1945, he circulated a petition at the Metallurgical Lab that the bomb not be used. To do so, the petition read, "was to bear the responsibility of opening the door to an era of devastation on an unimaginable scale."[3]

Szilard even attempted to galvanize support from scientists in other Manhattan laboratories. He wrote to Edward Teller in Los Alamos seeking support to stop the combat use of the bomb. Teller first consulted Oppenheimer, who replied that it was inappropriate for scientists to use their prestige as a basis for political statements. Teller replied to Szilard in an unusual letter in which he said that he had no hopes of clearing his conscience. "The things we are working on are so terrible that no amount of protesting or fiddling with politics will save our souls." Then he added:

But I am not really convinced of your objections [to the use of the bomb]. I do not feel that there is any chance to outlaw any one weapon. If we have a slim chance of survival, it lies in the possibility to get rid of war. The more decisive the weapon is, the more severely it will be used in any real conflict and no agreements will help.[4]

General Groves was furious when he heard of Szilard's petition. He considered it not only meddling, but something akin to disloyalty. Other scientists at the Met Lab agreed with Groves and submitted counterpetitions of their own.

Stimson's Scientific Panel met again in Los Alamos to discuss postwar policies. On June 16, the panel prepared three reports and sent them to Washington. Their recommendations to Stimson were broad and served both immediate and long-range projections: the panel urged the government to spend a billion dollars or more in the coming year on a diverse research program; it also urged that the Manhattan District continue its operation for another year or so, until a government agency could be created by Congress. The final report focused on the use of the bomb and reflected the fears and thinking of many scientists. The bomb should be used in a manner that would encourage nations to support a common agreement on the bomb's future use and on a system to control it. The report even stated that opinion differed among scientists, but it concluded with the belief that the war against Japan must be ended quickly and with the least cost in American lives.

The Chicago discussions of May 31 reemerged in the report: there was the possible use of the bomb in a controlled demonstration for Japan and the world to witness. This alternative offered the hope that such a demonstration would prevent American "dirty hands" and promote world cooperation. Interestingly, Oppenheimer, as well as Fermi, Lawrence, and Arthur Compton, were in favor of military use with one caveat: that such use be made only with the preknowledge of America's major allies, such as France, England, China, and Russia. They even reaffirmed this position in a meeting in Washington in early June. Essentially, they urged the use of the bomb against an enemy that would not be warned.

Unlike other laboratories, Los Alamos was far too busy trying to deliver Fat Man and Little Boy to engage in petitions. Only after Trinity were scientists freed from their demanding schedules to contemplate the question of the bomb's use against Japan. For most at Los Alamos the question was moot. Their energies for nearly three years had been directed at producing a weapon with the implicit assumption that it would be used. Trinity was perhaps sufficient reward for many. For others, however, only the bomb's use against the enemy would serve as a termination of their involvement. Oppenheimer was not insensitive to his people. He argued as strongly as he could for careful use of the atomic bomb, but in the end he moved to urge its use against Japan in an effort to conclude the war with dispatch. He spoke as he felt, and had the support of his colleagues at Los Alamos.

Ultimately Hiroshima revealed the futility of discussion. Control over the weapon's use had clearly moved into other hands. The sudden success did have the effect of germinating a more somber realization of the larger world, and many scientists felt a sense of involvement in the emerging postwar world for the first time. Even Stimson, in a press conference following the Hiroshima and Nagasaki drops, revealed a concern and deeper emotion: The bomb, he said, "is so terrific that the responsibility of its possession and its use must weigh heavier on our minds and our hearts."[5]

The Japanese surrender seemed, for the moment at least, to justify the use of the bomb. At Los Alamos, other issues began to surface the uncertainty of the Laboratory's future. One concern was the continuation of security and isolation beyond the war's end. John Manley, whose work with Oppenheimer in 1943 had laid the original plans for the Laboratory, wrote General Groves in disgust. Manley stressed the point that scientists at Los Alamos resented the military's failure to involve scientific staff in policy decisions and that such isolation would only serve to force the best men away from the Laboratory. At the Chicago

Laboratory, scientists again submitted a petition to President Truman requesting that he immediately share the bomb's secrets with other nations. Small groups were formed at Los Alamos, and one asked Oppenheimer's help in securing Army approval for release of a brief paper outlining the importance of the atomic bomb in world affairs.

Groves had his hands full. The end of the war suddenly released a flood of interests and fears on all fronts of the Manhattan District. Truman was not without his own concerns. Acting on advice from several groups, he called for legislation that would put atomic energy into the hands of a permanent civilian authority. Early legislation, called the May-Johnson Bill, brought before the public acrimonious division between scientists. Oppenheimer, among others, was solicited to support the legislation during hearings. Their discussions were hampered by security restrictions that pleased few except the military, who had come to view the atomic bomb as their exclusive chattel. Oppenheimer was even persuaded to recruit the support of Fermi and Lawrence in hearings before Congress. In his own appearance before a Congressional committee, Oppenheimer fielded questions from Congressman May with assurance and aplomb. He openly refused to support his colleagues' fears of military domination over the proposed civilian agency. Although other legislation was finally passed, the hearings served to widen the differences among the scientific camps.

But even as the debate quickened in Washington, Oppenheimer was left with responsibility for work at Los Alamos. The Japanese surrender eased many of the pressures on the Laboratory but created new ones: job placement, work continuity and, in general, the return to civilian life. Oppenheimer had to bear the brunt of these concerns and the demands on his time were staggering. Appearances in Washington constantly interfered with his ministrations to the Laboratory.

Much of the problem in Los Alamos was a result of more free time than work. There still was much to be done, of course, but hardly enough to keep everyone busy, and certainly not in an atmosphere of urgency. Oppenheimer was hedging on new scopes of work. The uncertainty of the Laboratory's future seemed to warrant caution in beginning new and expensive research and development. The best bet was simply to perfect or refine the products at hand.

Parsons, Kistiakowsky, and others agreed with Oppenheimer that the present Fat Man was prototypic at best. The uranium tamper was too large and inefficient and the explosive lenses were fragile. Research had already suggested several new shapes. Even the bomb's electronic system was crude. With approval from Groves, Oppenheimer put his men to refining the Fat Man components. The word was passed to group and division leaders to keep everyone busy, even if it meant undertaking activities of little value. The machinations in Washington left many in Los Alamos with a feeling that Groves and Washington officials cared little about their work. Kistiakowsky issued a lengthy memorandum to his group and section leaders titled "What to Do Now." Kistiakowsky was careful to add that the "What" might well be changed "on instructions from Washington."

Without a crash program under way the Laboratory found itself for the first time discouraging overtime, and staff members and their families were encouraged to take accumulated leaves and vacations. Kistiakowsky, among others, also discouraged individuals from leaving the Laboratory immediately. Several months at least would be needed for Los Alamos to be—as he said—"liquidated or reorganized," and at that time, staff members could leave without harm to the project. The senior men followed Oppenheimer's lead. He had begun to appear stronger in his belief that Los Alamos would survive in one form or another, or perhaps in another location. Everyone was assured that sudden notices of termination would not be given. In mid-August, the Laboratory was urging its staff to undertake a more thorough study of "fundamentals" of nuclear explosions and a return to the study of basic properties and dynamics of nuclear materials.

Partly as insurance against the sudden departure of key men, Oppenheimer organized a Los Alamos "Encyclopedia" as both a history and a documentation of the Laboratory's work. Hans Bethe was put in charge. He, in turn, asked men like Robert Wilson, Kennedy, Cyril Smith, Froman, and others to write major volumes in their respective scientific or technical areas. Each volume would then have chapters written by staff members. The Encyclopedia, as Bethe saw it, had two purposes: to make available to scientists the results and methods developed at Los Alamos, as well as put on record the techniques of making an atomic bomb. He added wryly that it might be "useful to our prospective successors."

The end of summer also brought an unexpected bonus: with the existence of Los Alamos revealed to the world, Laboratory staff suddenly received job offers from many institutions. Senior men surely had their pick, but even junior scientists and technicians found themselves with laboratories or universities to go to. The dire predictions of the previous summer had not come true. The war had drawn most young men into active service, and many with intentions of college work had simply been unable to start or perhaps complete degrees. Those with deferments at Los Alamos, or the few lucky service men with college degrees who were stationed on the Hill, suddenly possessed critically needed skills and experience for hungry laboratories and industry. Oppenheimer's urgings to stay on became more vocal.

Not everyone had a job, of course, but there were other reasons to leave. The end of the war magnified the isolation and uncomfortable life-style that many men and their families had been forced to accept at Los Alamos. The secrecy and poor housing were major grievances throughout the war, and they became intolerable for many without the urgency of war to motivate them. The many problems over fair salaries had never been fully solved. Isolation was also a factor. Many families had not seen their relatives and friends for three years. There was a perpetual shortage of water and many food items, and many had simply grown bored with the meager amenities allowed them in the area. With so much of the country returning to normal, the disenchantment with Los Alamos grew daily. Oppenheimer and his leaders were hard pressed to keep the Laboratory functioning against these odds. By October, the staff had fallen to an all-time low of 1,000 employees.

General Groves had his own problems to contend with. Between the return to normality and the shifting currents in Washington, Groves felt an additional strain in keeping his Manhattan apparatus running. The vast production facilities at Oak Ridge and Hanford were still necessary as suppliers of Uranium 235 and Plutonium 239. Groves, however, was forced to trim their operations by eliminating extra staff and production operations that had been created at a time when huge sums of money could be expended to meet the urgent delivery deadlines. The liquid-thermal diffusion plant at Oak Ridge was closed. A few weeks later, the Alpha Plant was shut down. Hanford technicians put themselves to perfecting operation procedures and to exploring ways to store plutonium nitrates.

Both Hanford and Oak Ridge had a single purpose: the production of critical material. Los Alamos was different in that it served both research and production needs. Groves was mindful of Oppenheimer's recommendation that Laboratory staff be permitted to continue research into more efficient bomb designs, as well as to eliminate unnecessary production chores such as uranium and plutonium purification and the manufacturing of explosive lenses. With the political fluctuations in Washington, however, Groves could not be sure what sort of government agency would emerge. For that reason, it seemed best to keep Los Alamos as it was. He had already permitted the Laboratory to take over a little-used airbase in Albuquerque called

Sandia Field. Groves approved of the use of Sandia for centralizing the assembly of atomic weapons and ordered the Wendover Field facilities in Utah to transfer to Albuquerque.

Oppenheimer also persisted in his request to resign and return to the University of California at Berkeley. Groves could be thankful for Oppenheimer's work as well as for his departure. His choice of Oppenheimer as director had been a good one, and success had vindicated the unusual action taken during his security clearance review. But Oppenheimer's departure would free him to lobby in Washington, and would permit Groves to replace him with someone more sensitive to military interests. While Oppenheimer had worked surprisingly well with Groves, he was still a scientist. To Groves, at least, the high drama over bomb control suggested the need for firmer hands.

Finding a replacement was more trouble than Groves expected. Many of the Laboratory's best men were scientists and were leaving to return to academic posts. Because Los Alamos still fell under Manhattan District control, Groves would have the final responsibility for selecting a new director. Groves was wise enough to ask Oppenheimer for recommendations and to sound out their compatibility with Los Alamos staff. But with the staff dwindling each day, Groves needed to act quickly.

Oppenheimer had his own personal thoughts on the matter of replacement. No doubt he considered several individuals and tentatively sought their reactions. He was too much a diplomat and politician not at least to offer the job to several men. Oppenheimer was aware, of course, that some of his candidates were returning to university life, and a few, perhaps, were approached and rejected the offer. High on Oppenheimer's list, and respected by Groves, was Norris Bradbury. Coming to Los Alamos as a Lieutenant Commander in the Naval Reserves, Bradbury had had an outstanding career at Pomona, the University of California, Massachusetts Institute of Technology, and Stanford. He had come in 1944 as a physicist as well as one of the Laboratory's few weapons experts. He had played major roles at Trinity and in Project Alberta and had flown in the observer airplane over Nagasaki. For Groves, he was a man sensitive to military interests with an impressive record at Los Alamos; for Oppenheimer he was a scientist, a natural leader with a fighter's heart, and someone who promised to take a strong hand in leading Los Alamos through the uncertain months ahead.

Oppenheimer was adamant on at least one point: Los Alamos must continue under the leadership of *one* director. At the last moment Groves balked and suggested to Oppenheimer that perhaps an interim director and a coordinator might be the best solution. Oppenheimer reacted strongly and placed an urgent call to Groves in

Washington to protest. Groves was not available, and Oppenheimer left a terse message with an assistant. There was only one job, he said, and that was as interim director *and* coordinator. His argument, Oppenheimer said, was acceptable to Los Alamos staff. Oppenheimer stood firm. He even went on to say to the startled secretary that when he and Bradbury came to Washington to meet with Groves, it would be with the assumption that there was but one job, and that if Bradbury was picked he would be picked "for the one job." Oppenheimer hung up and the secretary noted on the call sheet that he had never heard Oppenheimer "feel so strongly" about any matter.[6]

No one could be sure of the Laboratory's future or how Bradbury would manage it. Groves and Oppenheimer struck a compromise: Bradbury was asked to serve as interim director for six months. To assure cooperation from remaining staff members, Groves announced that he expected Oppenheimer and division leaders to choose a permanent director among themselves. Groves in fact visited Los Alamos on September 18. Speaking to Oppenheimer, Bradbury, and division authorities, he reviewed the May-Johnson Bill then before Congress, as well as his perceptions of the Laboratory's future.

Groves was unable to say what turns the Laboratory might take, except that it would retain its focus as a center for weapons research with some production of key bomb parts. The staff could expect this line of work for at least a few years. Oppenheimer had already told his leaders in late August that Los Alamos would continue to manufacture and stockpile weapons for at least some time—at least until Congress passed legislation and appropriated funds for further research. Groves also announced his appointment of Bradbury for a period of six months. Oppenheimer, they were told, was due to resign on October 16.

His resignation left Bradbury with a monstrous task: to hold together what staff he could, negotiate the Laboratory's future, and take on the personal responsibility for continuing America's domination of atomic weapons. It was not to be easy. Much of the spirit promoted by the war and the threat of a Nazi bomb was gone. The harsh realities of life in Los Alamos remained. Good men, even key men perhaps, had left or were preparing to leave. Bradbury could feel that he was being left with what James Tuck called the "second team."[7] There was also something less than agreement between scientists on what to do: Pursue the hydrogen bomb? Perfect more powerful Fat Men? Diversify the Laboratory's work into more areas of nuclear research, including those not related to weapons? By September, Oppenheimer seemed remote and disengaged from the Laboratory, as if his leadership had ended and his responsibility had evolved into more ethereal realms. Groves was busy in Washington defending his Manhattan District and battling Congress over legislation. Bush and Conant were equally distant and involved in arranging atomic energy in perspective before its new world audience. Truman was under pressure from all sides to regard the new weapon as leverage against a Soviet Union increasingly perceived as hostile. Planning was left to Bradbury.

Bradbury called his men together and began to assess Los Alamos. No one doubted that the government would continue to support bomb research, and that in all likelihood the research would be controlled by a new government agency or commission. Bradbury was sensitive to weaknesses inherent in some of the proposed legislation. It was possible that overcautious security would restrict the number of good men that could be attracted to Los Alamos. It was even possible that the new agency leaders would be mediocre and insensitive to research needs. There was little choice, however, but to plan for work considered "ideal" by Los Alamos standards.

"Ideal" meant work favoring scientific interests while being consistent with postwar needs for stockpiling nuclear weapons. Bradbury—with Oppenheimer present—broke his assessment to an anxious staff on October 1. He emphasized that weapons to be stockpiled were not necessarily intended for use, but instead gave the nation the bargaining strength it needed to bring about world agreement. "To weaken the nation's bargaining power," he said, "during the Administration's attempt to bring about international cooperation would be suicidal." He hoped that weapons emphasis would decrease over the years, and said, almost as if to reassure himself, that "we are not a warring nation; the mere possession of weapons does not bring about war."[8]

As the plan was unveiled, it became clear that three elements were called for. The Laboratory would create the most suitable project possible. Moreover, the Laboratory would not discontinue weapons research until it was clearly possible—by legislation probably—to do so. And finally, the size of the Laboratory would be decreased to accommodate life on the mesa on a civilian basis.

Even a smaller, planned project would require good men. This was a crucial element in Bradbury's analysis of the Laboratory's condition. With senior men and trained scientists leaving daily, there were good men left, but they were generally without management experience. Additionally, the trained enlisted men provided by the Army were leaving to relocate or to return to civilian life. It was critical, Bradbury said, that each division leader prepare a plan of solid research; one that was of intellectual interest and that would permit good men to stay or come. Next, Bradbury called for a massive overhaul of the

Laboratory's salary system, which had clumsily operated on a no-loss, no-gain principle and on no small amount of patriotism.

While money would not be the only factor in retaining men or attracting them, it was imperative to have a fair system. Oppenheimer and Groves had never put their house in good financial order. Bradbury was left to install a new policy that would be capable of competing with other laboratories and with universities. Where possible, the Laboratory would now try to make counteroffers as a means of keeping on-board staff. And almost as important, a new housing policy would be created that would make available fair housing to both professional and technical staff. Everyone required housing, and the size and operation of the Laboratory was fixed by its available housing.

Bradbury referred lightly to the "General," and indicated that it was unlikely that Groves would permit the construction of new housing until legislation was pushed through Congress. As a result, Los Alamos as both a community and a laboratory would have to restrict its operations to accommodate individuals and families that could be housed. He called for each division to assess its work and to plan for a slowdown, reducing the work by two thirds. "It is curious," he offered, "that the activity of the mesa should be dictated by its housing."

And what of the work to be done? Bradbury asked that at least fifteen Fat Men be stockpiled. These bombs would require modifications along the lines proposed by Oppenheimer: new methods of fuzing, detonation, and packaging. A program would begin seeking more efficient and more powerful weapons based on altered implosion techniques. Some of this work was already underway, and would not be stopped. It would be a few years—Bradbury speculated about three—before production of weapons could stop. Until that time, however, the goal was a weapon that would "insure peace."

Perhaps most exciting, Bradbury called for more tests like Trinity that would serve several purposes. For Los Alamos, these purposes were the catastrophic forces that unleashed nature's rawest elements. These forces were the obvious target for scientific study. Likening the bomb to cancer, Bradbury said that such a disease must be studied, although one did not expect or want to contract it. Perhaps, he argued—and he spoke for many men at Los Alamos—the occasional demonstration of an atomic bomb would have a psychological effect on the world; an effect that properly witnessed might urge cooperation for control. And more lightly, he suggested that another Trinity might even be fun.

There was still the nagging question of the Super. Was such a bomb feasible, and could it realistically be built? Bradbury and Oppenheimer were alike in their views: the question presented many interesting scientific problems to study, no matter how terrifying. Bradbury considered the word "feasible" to be duplicitous. It encompassed everything from laboratory experiments to actual bomb construction. While he considered construction of the Super improbable at the moment, it was a question waiting for an answer.

Teller continued to push for an acceleration of work on the Super. He approached Bradbury with his own concerns, but was unable to get a commitment from the new director. Teller then approached Oppenheimer and tried to enlist his support for the new bomb. Oppenheimer appeared unwilling to use his influence to promote such research.[9] Teller sensed quickly that his urgings would need a new audience in other quarters. At the moment, there was no way to foretell the events of the next few years, when Teller would rise to prominence and, with other scientists and politicians, argue successfully for the Super's construction.

For Oppenheimer, who sat quietly in the room listening to Bradbury, there was little to suggest that he would become the center of a vicious storm over just such a weapon. Nor could he predict that his private life and philosophy would be revealed and dissected by men far less honorable than himself. For the moment, Bradbury's careful, sometimes light, presentation had a salutary effect on the Laboratory's assembled leaders. There seemed to be hope for continuing their work and for keeping their staffs together. Even life on the Hill promised improvement.

Oppenheimer could feel a mixed emotion: satisfaction at the job done, tempered with an uneasiness for the future. He was leaving Los Alamos—a place he had first seen when it was little more than an empty mesa and a private school for boys. He was leaving men he himself had brought to Los Alamos. The work was now in their hands. It had been exciting, and when once he had been asked why scientists—including himself—had been attracted to such a weapon of destruction, he had replied, "It was technically sweet."

Bradbury, too, could not foretell how truly difficult his task would be. There was no way of knowing that Washington politics would delay creation of an agency for atomic energy for many months. More importantly, there was no way of predicting the substantial and pervasive effect of politicians on the course of American nuclear research. In the meantime, more staff left Los Alamos. The primary water main froze in February and left residents without running water for weeks. And more than a year would pass in which the Laboratory's future would swing back and forth in uncertainty.

For Oppenheimer's departure, however, festivities were planned. Groves arrived with a cadre of important officials to present to

Los Alamos and to Oppenheimer a Certificate of Appreciation from the Secretary of War. October 16 found Los Alamos with bright, cool weather. The front of Fuller Lodge was decorated with flags and colorful bunting. From a low platform, Groves presented the certificate to Oppenheimer and made a brief speech acknowledging the important work completed by the men and women of Los Alamos. It was as much a ceremony to honor the man as his work. More quietly, it transferred the mantle to Norris Bradbury.

Later that night at a large party in his honor, Oppenheimer was again surrounded by old friends and colleagues. He was given a hand-carved wooden chest that staff members had purchased from the La Fonda Hotel in Santa Fe. The moment was only superficially joyous; most men and women knew that Los Alamos was losing something important. It was more than Oppenheimer's leaving, although it was fixed with his departure. The community felt an unknown change at hand. What had been so eagerly, if not quixotically, done in war, was now to be done in peace. And yet, as Oppenheimer would have it, the atmosphere of the Hill was irrevocably different.

Oppenheimer had brought them together in a great undertaking. Monumental discoveries had been made, and new forces had emerged to contend with. And this new knowledge seemed both promising and ominous. Oppenheimer had said it best in his speech earlier in the afternoon. Accepting the citation from Groves, he spoke in a low, careful voice. He thanked Groves on behalf of the Laboratory's men and women, and expressed the hope that everyone could look at the scroll in the future with pride. Pausing, he said that, "if atomic bombs are to be added to the arsenals of a warring world, or to the arsenals of nations preparing for war, then the time will come when mankind will curse the names of Los Alamos and Hiroshima." Oppenheimer paused again and added a note of hope.

The peoples of this world must unite, or they will perish. This war, that has ravaged so much of the earth, has written these words. The atomic bomb has spelled them out for all men to understand. Other men have spoken them, in other times, of other wars, or other weapons. They have not prevailed. There are some, misled by a false sense of human history, who hold that they will not prevail today. It is not for us to believe that. By our works we are committed, committed to a world united, before this common peril, in law, and in humanity.[10]

In the end, this was the judgment Oppenheimer set for himself and for Los Alamos.

PART THREE
RETRO-SPECTION

PHYSICISTS HAVE KNOWN SIN...
J. Robert Oppenheimer

17. RETRO-SPECTION: 1978

Los Alamos and the world are very different now. It has been over three decades since Leslie Groves and Robert Oppenheimer surveyed the Los Alamos Boys' School as a possible site for their Laboratory. Over thirty years have passed since Trinity and Hiroshima and Nagasaki.

These years have taken their toll and woven their changes. Los Alamos is no longer a collection of wooden Army buildings but a permanent city of 13,000 residents. The principal actors—Oppenheimer, Groves, Fermi, Parsons—are dead, and others grow old, and second and third generations of men live on the mesa. The world is different, too. The arsenals that Oppenheimer warned against have become a reality, and the size and power of these weapons dwarf Trinity and surpass even the most fanciful thinking of the early days.

The legacy is mixed. Los Alamos no doubt will always be remembered for its weapons: Fat Man and Little Boy have survived far beyond Trinity and Japan. Los Alamos will also serve as an example of the conflict between the scientist and society. For some, Los Alamos has even come to be regarded as the tragedy of the age, for just when science could have put its powers to good, it turned them to destruction.

And yet there is another dimension. There is no doubt that for those years during the war Los Alamos was an exceptional place. It was unique, most importantly because of the atmosphere created and fostered by men like Robert Oppenheimer. As he and his colleagues came to the Hill and invited others to join them, there was a continuation of the university spirit, the special relationship between the teacher and his students. Science in the 1930's was, after all, small and familial; professors acted like friends, surrogate fathers, and mentors. Young men coming to Los Alamos felt much the same spirit; they worked and were treated as members of the team. Young and old worked side by side and everyone felt the sense of intellectual challenge. Men came not so much to build a bomb as to be a part of that spirit.

Los Alamos was also special because of its participants. Truly gifted men of all ages came. There were ordinary men as well, but always there were giants: Oppenheimer, von Neumann, Fermi, Bohr, Bethe, and others. There were less prominent men whose careers were to coalesce at Los Alamos and who would leave to become leaders elsewhere: Kistiakowsky, John Williams, Bacher, Alvarez, Hornig, Bradbury, and dozens more. And there were junior scientists—hundreds of bright and energetic men—who grew up in Los Alamos and stayed, or left to take positions at universities and laboratories elsewhere.

There was the work, the long days and nights spent in unraveling mysteries. There was the setting: the majestic Jemez Mountains cut into the many-fingered plateau with its mesas and lonely canyons. There was the life of isolation, softened by companionship. There was a sense of family and common interest and hardships shared. There was the cosmopolitan mix of Americans and Europeans working together, and laboratories and parties where several languages were spoken. There was the newness set against the hollowed cliff dwellings and ancient Indian pueblos.

Their creation—the bomb—is a moral burden for very few. And of guilt? Little can be said of something so personal to the individual. There is, at least, very little sense of collective anguish. "It was war," one said. "Someone else would have made it," said another. Almost everyone who was there agrees that Los Alamos helped to bring the scientist, perhaps for the first time, into confrontation with politics.

The "special feeling" shared by Los Alamos alumni is more than just having worked together in isolation and in war. More at heart is what Alice Kimball Smith describes as the "dilemma of intellectuals caught in a sudden shift of values which they themselves helped to produce."[1] For others, it was more difficult. As Oppenheimer was to say, "In some sort of crude sense which no vulgarity, no humor, no overstatement can quite extinguish, the physicists have known sin; and this is a knowledge which they cannot lose."[2]

Historians and demagogues now debate the legacy of Los Alamos. Was the bomb really necessary? Would the war have ended without Hiroshima and Nagasaki? Like all great historical questions, there will never be any final resolution.

Of all the individuals who came to Los Alamos, certainly the most powerful and enigmatic was Robert Oppenheimer. He was clearly a man of brilliance and charm, and less obviously—in the manner of an ancient Greek character—a man of arrogance.

After the war and his resignation from Los Alamos, Oppenheimer returned with his family to Berkeley. He stayed only briefly, and continued his journey as elder statesman of atomic energy. His quick mind and insights were sought in all quarters of government; his sharp tongue brought him both admirers and enemies. Acting with many of his colleagues, he urged a cautious and planned exploration of the new weapon against the unreasonable proliferation of bigger and more powerful ones. He sought to hold back development of the hydrogen bomb until agreement with the Soviet Union could be reached to limit nuclear development.

The revelation of the Russian atomic bomb in 1949 ended hope of international agreement, and Oppenheimer found his influence increasingly mitigated in the rising flood of anticommunist hysteria. The newly created Atomic Energy Commission dropped Oppenheimer from its General Advisory Committee in 1952. Two years later, in 1954, with only months until his consultant contract expired, President Dwight D. Eisenhower and the Atomic Energy Commission chose to raise the matter of Oppenheimer's "security clearance." In question was Oppenheimer's loyalty to the United States and his risk to national security.

In a painful "hearing" before AEC-selected representatives, Oppenheimer was brutally attacked on, among other things, his past association with Communist groups in the 1930's and on the conduct of his personal life. At the heart of the hearing was Oppenheimer's actions during the late 1940's to stall development of the Super until all chances for international control had been exhausted. By 1954, both the United States and the Soviet Union possessed hydrogen weapons. Oppenheimer's activities were viewed as retarding American domination of weapons technology. Perhaps, as Oppenheimer himself finally was to say, he had been "an idiot" in his behavior; perhaps he had struck one too many Washington personalities with his arrogance.

Paradoxically, the AEC hearing found Oppenheimer loyal but nonetheless a "security risk." As a result, his "security clearance" was revoked and atomic secrets—many of which he had helped to discover—were denied to him. Later he was offered, and accepted, the directorship of Princeton University's Institute for Advanced Study. His years at the Institute were ones of solitude and reflection. The years helped to ease the pain of the hearing and to remove him even more from public scrutiny. He remained a villain to some and a hero to others.[2]

In 1963, President John Kennedy sensed the great wrong and awarded Oppenheimer the newly created Fermi Award for scientific achievement and contributions to humanity. Kennedy announced the award on the morning of his assassination in Dallas, Texas. Lyndon Johnson, in a strong act of courage, insisted on personally awarding the Fermi prize to Oppenheimer two weeks later. Robert Oppenheimer died quietly in February 1967. For the man, at least, the storm had finally abated.

Others from Los Alamos fared differently. Edward Teller experienced perhaps the most meteoric change. With the end of the war, Teller stayed on the Laboratory's staff for nearly a year. His preoccupation was with pushing ahead on development of the hydrogen bomb. Increasingly he experienced the division between his colleagues on further weapons development. Teller quickly became the major proponent of big weapons in the nation. In Washington, he was able to exploit a growing circle of friends and political leaders who similarly favored expansion of America's arsenal.

After a brief return to the University of Chicago, Teller was successful in arguing for the creation of a new weapons laboratory to pursue alternative hydrogen bomb designs. In 1952, along with some staff members from Los Alamos, Teller created the Lawrence Livermore Laboratory in northern California. Although he was its director only briefly—from 1958 to 1961—the new laboratory was useful as a center for his work on weapons, as well as a base from which to operate politically. More than any other individual, Teller was able to prevent the inclusion of a ban on underground testing in the 1963 Nuclear Test Ban Treaty. He continues today to exert influence on the development of new weapons and on the course of American nuclear strategy.

Enrico Fermi left Los Alamos at the end of the war and returned to the University of Chicago. While initially opposing the development of the Super, he ceased to argue against it after President

Truman's decision to proceed. He served as a consultant to Los Alamos and to Livermore. Fermi died of cancer in 1954.

John von Neumann similarly worked on the Super's development and became a top-level scientific advisor to the Defense Department. Von Neumann also served as a member of the Atomic Energy Commission's General Advisory Committee until his death in 1957.

Luis Alvarez left Los Alamos, eventually to join the Lawrence Berkeley Laboratory. Like Teller, he was an early supporter of hydrogen bomb development and served as a consultant to Livermore and Los Alamos. In 1958, he won the Nobel Prize for his work in physics.

Hans Bethe returned to his post as Professor of Physics at Cornell University. While he opposed development of the Super, he worked on its research with the hope that the bomb would not work. Bethe served on numerous official committees and acted as a consultant to Presidents Kennedy and Eisenhower. In 1967, he received the Nobel Prize for his youthful work on the thermonuclear properties of stars. Bethe has a vacation house in a quiet valley near Los Alamos.

John Manley spent most of his career after the war at the Laboratory. He is retired now and lives below Los Alamos near Española.

Norris Bradbury, who became the Laboratory's "interim director" in 1945, remained as leader through the difficult postwar years and until his retirement in 1970. Throughout his years as director, Bradbury walked the difficult line between promoting further weapons development and arguing for a sane arms control policy. He too retired in Los Alamos.

Leslie Groves continued as czar of the vast Manhattan District until Congress created the Atomic Energy Commission in 1947. After the transfer of the many MED projects to the new agency, including the Los Alamos Laboratory, General Groves continued briefly in the U.S. Army and then retired. He accepted an executive position in industry and began, in the early 1960's, to write his memoirs. Commandeering a private office in the National Archives Building, Groves kept researchers and secretaries busy with his orders. He died in 1973.

Two alumni of Los Alamos have special historical legacies. Both Klaus Fuchs and David Greenglass were eventually arrested and convicted for their wartime activities. In 1950, the British Government arrested Fuchs and tried him, not for high treason—for which the penalty was death—but for violation of the Official Secrets Act. Because Fuchs passed secret information to a wartime ally, instead of an enemy, he received only a fourteen-year sentence. After his release, Fuchs went to East Germany to live and again took up physics research.

Acting on information from the British, the Federal Bureau of Investigation was able to uncover the work of David Greenglass and Julius and Ethel Rosenberg. In a 1951 trial that lasted three weeks, the Rosenbergs were found guilty, given the death sentence, and were executed. Greenglass collaborated as a state's witness and was treated more leniently; he was sentenced to 15 years in a federal prison. He now lives quietly in the United States.

Los Alamos today is an open city. The gates were removed in 1957, and in 1962 John Kennedy signed legislation that returned to private hands most of Pajarito that had been owned by the government. The small city is still largely scientific, and children play with only the tales and shadows of the war around them. There are veterans from the early days who now occupy most of the present Laboratory's management. Of the many who left at the end of the war, a few have returned to work or to settle in Los Alamos or in nearby communities.

The Laboratory still receives hundreds of millions of dollars each year from the government. Half goes for weapons research, but now funds are also spent on new challenges and frontiers, like solar and fusion energy. The old Laboratory buildings have been torn down and replaced with permanent concrete structures. The Technical Area has been replaced by a motel, restaurants, and a new community center. Only Ashley Pond survives because of the indignation of local citizens, who fought to protect something of the past and to combat the modern penchant for replacing everything flat with concrete and asphalt.

New housing areas have flooded into nearby mesas, with premium prices still paid for World War II apartments. There are preferable and not-so-preferable areas of Los Alamos to live in. Only occasionally are residents reminded of the pioneering 1940's. Builders of a new motel in 1976 dug a foundation and discovered the area highly contaminated by plutonium and uranium residue from the old Technical Area.

Two hundred miles away, at Trinity, there is silence; the desert has reclaimed its own. The great crater at Ground Zero is little more than a shallow saucer in the vast floor of the New Mexico desert. The crystalline bits of fused green sand from the blast have been all but removed by the annual carloads of tourists who make the journey every October to see the birthplace of the atomic age. Base camp is gone, but for the sagging shapes of two wooden buildings. The McDonald ranch house stands, but shows signs of retreating into history with each wind and sandstorm. Scattered on the desert are still long strands of black wire that once ran from men and equipment straight to Ground Zero. The desert has outlived man's assault for those brief moments in 1945.

Twenty-five hundred years ago, the same Greeks who invented the concept of the atom also believed that it was the titan Prometheus who first gave fire to mankind as a gift. But because fire was the exclusive property of the gods, Prometheus was punished for his sacrilegious act by being chained to a rocky pinnacle for his body to be gnawed eternally by eagles.
    It will be ironic if Oppenheimer and Los Alamos suffer a similar fate in history. They, too, stole the power of the gods on man's behalf.

NOTES

ABBREVIATIONS

    LASL   Los Alamos Scientific Laboratory
                 Los Alamos, New Mexico

    MED   Manhattan Engineering District
                 National Archives, Modern Military Branch,
                 Washington, D.C.

Chapter 2: Origins

1. Early negotiations concerning acquisition of the Los Alamos area are covered in Volume I of the Manhattan District History, now partially declassified, in the collections of both LASL and MED. LASL has recently published this first volume as *Manhattan District History, Nonscientific Aspects of Los Alamos Project Y, 1942 through 1946*, Publication LA-5200, March 1973.

Chapter 3: World in Change

1. Richard Hewlett and Oscar E. Anderson, Jr., *A History of the United States Atomic Energy Commission, Volume I, the New World 1939–1946*. University Park, Pennsylvania: Pennsylvania State University Press, 1962, p. 17. Hewlett

and Anderson's thorough study of the origins and activities of the Manhattan District form a basic research tool for historians, especially those without access to classified materials.
2. Ibid., p.20.
3. Ibid., p. 34.
4. Ibid., pp. 37–38.
5. Ibid., p. 49.
6. Ibid., p. 55.

Chapter 4: The New Alliance

1. Hewlett, p. 73.
2. Interview with John Manley, April 1975.
3. U.S. Atomic Energy Commission, *The First Reactor*. Washington, D.C.: no date.
4. Hewlett, p. 115.
5. U.S. Atomic Energy Commission, *In the Matter of J. Robert Oppenheimer*. Washington, D.C.: Government Printing Office, 1954, p. 12.
6. Davis, Nuel Pharr, *Lawrence & Oppenheimer*. New York: Simon and Schuster, Inc., 1968, pp. 141–145.
7. Robert Oppenheimer's prewar activities are discussed in the AEC's *In the Matter of J. Robert Oppenheimer* and in Davis' *Lawrence & Oppenheimer*.
8. U.S. Atomic Energy Commission, *In the Matter of J. Robert Oppenheimer*, p. 170.
9. Davis, pp. 187–190.

Chapter 5: Laboratory on a Hill

1. Contract between the United States of America and the University of California, dated April 20, 1943. LASL files.
2. Letter from Richard C. Tolman to Brigadier General Leslie Groves, March 20, 1943. LASL and MED files.
3. Los Alamos Scientific Laboratory, *Manhattan District History Project Y: The Los Alamos Project*. Report written by David Hawkins and published as LAMS-2532 (Volume I), 1961.
4. Memorandum from Richard C. Tolman entitled *Memorandum on Los Alamos Project as of March 1943*. LASL files. This memorandum forms one of the basic theoretical documents of the Laboratory. Copies were made available to arriving scientists in 1943 as background reading.
5. Hewlett, p. 236.

Chapter 6: The Actors

1. Memorandum from Robert Oppenheimer to General Leslie Groves, November 20, 1942. LASL and MED files.

Chapter 7: Organization, 1943

1. Contract between United States of America and University of California, April 20, 1943. LASL files.
2. U.S. Atomic Energy Commission, *In the Matter of J. Robert Oppenheimer*, p. 11.

3. Ibid., p. 12.
4. Groves, Leslie R., *Now It Can Be Told*. New York: Harper & Row, 1962, pp. 254–256.
5. Ibid.
6. Minutes of the Governing Board, May 31, 1943. LASL files.
7. Hewlett, pp. 238–239. Although in retrospect the issue of "compartmentalization" seems unimportant, it was crucial—as most participants agree—to the quick development of weapons at Los Alamos through the open interchange of ideas between all scientists.
8. Letter from General Leslie Groves to Commanding General, Services of Supply, February 27, 1943. MED files.
9. Letters from General Leslie Groves to Robert Oppenheimer, one on June 18, 1945, and the other, undated, but of the same period.

Chapter 8: The First Year, 1943—1944

1. Hewlett, p. 243.
2. Minutes of the Advisory Board, August 17, 1944. LASL files.
3. Minutes of the Governing Board, June 17, 1943. LASL files.
4. Minutes of the Governing Board, September 23, 1944. LASL files.
5. Minutes of the Governing Board, November 4, 1943. LASL files.
6. Memorandum from George Kistiakowsky to Robert Oppenheimer and William Parsons, June 13, 1944. LASL files.
7. Memorandum from Robert Oppenheimer to Seth Neddermeyer, June 15, 1944. Library of Congress, Oppenheimer collection.
8. Minutes of the Advisory Board, July 20, 1944. LASL files.

Chapter 9: Life on the Hill

1. Brode, Bernice, "Tales of Los Alamos," *LASL Community News*, June 1960.
2. Letter from Robert Oppenheimer to General Leslie Groves, June 21, 1943. LASL files.
3. Los Alamos Scientific Laboratory, "Second Memorandum on the Los Alamos Project," no date. LASL files.
4. Minutes of the Governing Board, October 28, 1943. LASL files.

Chapter 10: The Second Year, 1944—1945, Part I

1. Hewlett, p. 251.
2. Minutes of the Governing Board, June 29, 1944. LASL files.
3. Minutes of the Advisory Board, August 17, 1944. LASL files.
4. Minutes of the Advisory Board, August 3, 1944. LASL files.
5. Minutes of the Advisory Board, July 20, 1944. LASL files.

Chapter 11: The Second Year, 1944—1945, Part II

1. Hewlett, p. 252.
2. Ibid.
3. Minutes of the Governing Board, June 17, 1944. LASL files.
4. Blumberg, Stanley A. and Gwinn Owens, *Energy and Conflict: The Life and Times of Edward Teller*. New York: G. P. Putnam's Sons, 1976, pp. 128–131.

Despite Bethe's telling of this story, Teller denies that he ever refused such a request.
5. Hewlett, p. 240.
6. Los Alamos Scientific Laboratory, *A Review of Criticality Accidents*. Written by William R. Stratton and published as LA-3611, September 1967.
7. Memorandum from Dr. Louis Hempelmann to Dr. David Dow, September 19, 1945. LASL files.
8. Memorandum from Darol Froman to Norris Bradbury, "Preliminary Report on the Accident in Pajarito Laboratory," May 21, 1946. LASL files.
9. Minutes of the Advisory Board, December 7, 1944. LASL files.
10. Hewlett, p. 374.
11. Memorandum from James F. Byrnes to Franklin D. Roosevelt, March 3, 1945. MED files.

Chapter 12: The Harvest

1. Letter from Robert Oppenheimer to General Leslie Groves, May 7, 1945. LASL and MED files.

Chapter 13: Homestretch

1. Minutes of the Governing Board, September 16, 1943. LASL files.
2. Letter from General Leslie Groves to General Dwight Eisenhower, March 22, 1944. MED files.
3. Pacific War Research Society, *The Day Man Lost: Hiroshima, 6 August 1945*. Palo Alto, California: Kodansha International/USA., 1972, pp. 65–220.
4. Letter from General Leslie Groves to Field Marshal Sir John Dill, Combined Chiefs of Staff, January 17, 1944. MED files.
5. Henry Stimson Diaries. Quoted by Martin J. Sherwin, in *A World Destroyed*. New York: Alfred A. Knopf, Inc., 1975, p. 5.
6. Sherwin, p. 136.
7. Sherwin, p. 198. Sherwin's well-researched work is the best study to date of the complicated "politics" of the atomic bomb during the war and immediate postwar years.
8. Bainbridge, Kenneth T., *Oral History File*, no date. LASL files.
9. Memorandum from Robert Oppenheimer to George Kistiakowsky, October 13, 1944. LASL files.
10. Histories of Los Alamos and the Trinity Test have frequently mistranslated "Jornado del Muerto" as "the journey of death." Fray Angelico Chavez sets this aside in his historical work, *My Penitente Land* (Albuquerque: University of New Mexico Press, 1974) as "the dead man's route."
11. Bainbridge.
12. Davis, p. 230.
13. Memorandum from James Tuck to Robert Oppenheimer, et al., June 30, 1944. LASL files.
14. Bethe, Hans and Robert Christy, "Memorandum on the Immediate Aftereffects of the Gadget," March 30, 1944. LASL files.
15. Memorandum from Kenneth Bainbridge to all members of the Trinity Project, March 21, 1945. LASL files.
16. Memorandum from Kenneth Bainbridge to Thomas Jones, May 2, 1945. LASL files.

17. Letter from Richard Tolman to General Leslie Groves, May 13, 1945. LASL files.

Chapter 14: Dawn

1. Los Alamos Scientific Laboratory, *Los Alamos: Beginning of an Era: 1943–1945*, no date.
2. Bainbridge, Kenneth, "A Foul and Awesome Display," *Bulletin of the Atomic Scientists*. May 1975.
3. Memorandum from Captain T. C. Jones to Major Claude C. Pierce, July 30, 1945. MED files.
4. Los Alamos Scientific Laboratory, *Los Alamos: Beginning of an Era: 1943–1945*. A more complete report on the Trinity test exists in LA-1012, entitled "Trinity." LASL files.
5. ———, *Manhattan District History Project Y*.
6. Laurence, William L., *Dawn Over Zero: The Story of the Atomic Bomb*. New York: Alfred A. Knopf, Inc., 1946, p. 4.
7. Los Alamos Scientific Laboratory, *Los Alamos: Beginning of an Era: 1943–1945*.
8. Ibid.
9. Notes, General Leslie Groves, from a meeting in Chicago, July 24, 1945. MED files.

Chapter 15: Japan

1. Interoffice Memorandum, Robert Oppenheimer, June 13, 1944. LASL files.
2. Letter from Robert Oppenheimer to General Leslie Groves, September 20, 1944. LASL and MED files.
3. Memorandum from Robert Oppenheimer to General Leslie Groves, December 15, 1944. LASL files.
4. Memorandum from Kenneth Bainbridge and George Kistiakowsky to William Parsons, July 17, 1945. LASL files.
5. Memorandum from Robert Oppenheimer to General Leslie Groves and William Parsons, July 23, 1945. LASL files.
6. Notes on Interim Meeting of Target Committee, May 2, 1945. MED files.
7. Quoted in Sherwin, p. 204.
8. Ibid., p. 205.
9. Hewlett, pp. 365 and 394.
10. Memo from General Thomas Farrell to General Leslie Groves, September 27, 1945, p. 2. MED files.
11. Ibid., p. 8.
12. Thomas, Gordon and Max Morgan Witts, *Enola Gay*. New York: Stein & Day, Publishers, 1977, p. 276.
13. Memorandum from Farrell to Groves, p. 9.
14. Memorandum from General Leslie Groves to Chief of Staff George Marshall, August 10, 1945. MED files.
15. Transcript, telephone conversation between General Leslie Groves and J. Robert Oppenheimer, August 6, 1945. MED files.

Chapter 16: Change of Guard

1. Minutes of the Administrative Board, December 7, 1944. LASL files.
2. Letter from J. Robert Oppenheimer to General Leslie Groves, May 7, 1944. LASL and MED files.
3. Hewlett, p. 399.
4. Letter from Edward Teller to Leo Szilard, July 2, 1945. MED files and quoted in Blumberg and Owens, pp. 156–157.
5. Hewlett, p. 416.
6. Transcript, telephone conversation between J. Robert Oppenheimer and General Leslie Groves, September 15, 1945. MED files.
7. Davis, p. 250.
8. Bradbury, Norris, "Notes on Talk Given by N.E. Bradbury at Coordinating Council, October 1, 1945," date October 8, 1945, LASL files.
9. Cited in Blumberg and Owens, pp. 186–187.
10. Los Alamos Scientific Laboratory, *Manhattan District History Project Y*, p. 294.

Chapter 17: Retrospect: 1978

1. Lewis, Richard and Jane Wilson, editors; *Alamogordo Plus Twenty-five Years*. New York: Viking Press, 1971, p. 46.
2. Quoted in Herbert York's *The Advisors: Oppenheimer, Teller and the Superbomb* (San Francisco: W. H. Freeman and Co., Publishers, 1976), p. 47.
3. The most comprehensive account of Robert Oppenheimer's hearing before the Atomic Energy Commission is in Phillip M. Stern's *The Oppenheimer Case: Security on Trial* (New York: Harper & Row, Publishers, 1969).

SOURCES & BIBLIOGRAPHY

The richest source of information on the wartime Laboratory exists within the files and records of the present Los Alamos Scientific Laboratory in Los Alamos, New Mexico. Unfortunately for the historian, no systematic attempt was made during World War II to "organize" records, and all printed material was considered classified information. Material of historical relevance, therefore, appears in a number of file categories. Major sources are as follows:

> • The Laboratory's *Technical Report Series,* which includes a substantial amount of administrative detail locked within technical reports. Many of the early documents are still classified.

> • The minutes of the Laboratory's Governing and Administrative boards which operated from 1943 to 1944 and contain much information on the day-to-day decisions and operations of the Laboratory.

> • The minutes of various *ad hoc* Laboratory committees created for special purposes: e.g., the "Cowpuncher" Committee and the Trinity Organization.

- The correspondence between Robert Oppenheimer and others and major Manhattan Project figures such as General Leslie Groves, Vannevar Bush, James B. Conant, and Richard Tolman.

- The internal Laboratory memoranda between individuals and administrative groups.

- The *Manhattan District History*, a multivolume series, which is still in part classified and exists in duplicate in the Manhattan District collection of the National Archives.

The National Archives in Washington, D.C., contain two major collections of interest to the study of Los Alamos. The first is the files of the *Office of Scientific Research and Development, S-1*, within the Industrial and Social Branch of the Archives. This collection contains the files and records of ORSD under the direction of Vannevar Bush and his assistant, James B. Conant. The more relevant collection, however, is the extensive holdings of the *Manhattan Engineering District* in the Modern Military Branch. This collection has three major components: the general records of the MED; a special assortment of material entitled "Top Secret–Special Interest to General Groves" and gathered for use by Groves in preparing his biography; and the files of the Secretary of War, Henry Stimson, and entitled the Harrison-Bundy Collection. Each of these minicollections contains material illuminating the role of Los Alamos during the war.

Two other collections contain useful information. The Energy Research and Development Administration (ERDA), formerly the Atomic Energy Commission, keeps historical documents, many of which are duplicates of those in the National Archives. At present these materials are under the control of ERDA's Division of Classification. The Library of Congress now contains the private papers of J. Robert Oppenheimer.

Two published books are of particular use to the history of the early Los Alamos Laboratory. Under the aegis of the Laboratory, David Hawkins wrote the *Manhattan District History, Project Y, The Los Alamos Project* (Los Alamos, New Mexico; Los Alamos Scientific Laboratory, LAMS-2532, 1961) in 1946, and a declassified version was not made public until 1961. Although it is largely a technical history, it does provide interesting insights into the many scientific developments that necessitated administrative action. For the historian, the more useful work is that by Richard G. Hewlett and Oscar E. Anderson, *The New World, 1939/1946, Volume I, A History of the United States Atomic Energy Commission* (University Park: The Pennsylvania State University, 1962). For this work, commissioned by the AEC, the authors were permitted full access to all records in multiple collections. An extensive footnoting system permits the interested reader to identify key documents and their location.

The following published books and documents are also relevant to the early years of the Los Alamos Scientific Laboratory and to the general history of the times and to the Los Alamos geography.

Blumberg, Stanley A. and Gwinn Owens, *Energy and Conflict: The Life and Times of Edward Teller*. New York: G. P. Putnam's Sons, 1976.

Brode, Bernice, *Tales of Los Alamos*. Appeared in the "Los Alamos Scientific Laboratory Community News," Los Alamos, New Mexico, June through August, 1960.

Brown, Anthony Cave and Charles B. MacDonald, *The Secret History of the Atomic Bomb*. New York: The Dial Press, 1977.

Bush, Vannevar, *Modern Arms and Free Men*. New York: Simon & Schuster, Inc., 1949.

Byrnes, James F., *Speaking Frankly*. New York: Harper & Row, Publishers, 1947.

Chevalier, Haakon, *Oppenheimer: The Story of a Friendship*. New York: George Braziller, Inc., 1965.

Church, Peggy Pond, *The House at Otowi Bridge*. Albuquerque:University of New Mexico Press, 1959.

Compton, Arthur H., *Atomic Quest: A Personal Narrative*. Princeton, N.J.: Oxford University Press, 1956.

Davis, Nuel Pharr, *Lawrence and Oppenheimer*. New York: Simon & Schuster, Inc., 1968.

Fergusson, Erna, *Our Southwest*. New York: Alfred A. Knopf, Inc., 1941.

Fermi, Laura, *Atoms in the Family*. Chicago: University of Chicago Press, 1954.

Gowing, Margaret, *Britain and Atomic Energy: 1939–1945*. New York: The Macmillan Company, 1964.

Groueff, Stephane, *Manhattan District*. New York, Little, Brown and Company, 1967.

Groves, Leslie R., *Now It Can Be Told*. New York, Harper & Row, Publishers, 1962.

Irving, David, *The German Atomic Bomb*. New York, Simon & Schuster, Inc., 1967.

Jette, Eleanor, *Inside Box 1633*. Los Alamos, N.M.: Los Alamos Historical Society, 1977.

Jungk, Robert, *Brighter Than a Thousand Suns*. New York: Harcourt Brace Jovanovich, Inc., 1958.

Knebel, Fletcher and Charles W. Bailey III, *No High Ground*. New York: Harper & Row, Publishers, 1960.

Lamont, Lansing, *Day of Trinity*. New York: Atheneum Publishers, 1965.

Lang, Daniel, *Early Tales of the Atomic Age*. Garden City, N.Y.: Doubleday & Company, Inc., 1948.

Laurence, William L., *Dawn Over Zero*. New York: Alfred A. Knopf, Inc., 1946.

Lewis, Richard S. and Jane Wilson, editors, *Alamogordo Plus Twenty-five Years*. New York: The Viking Press, Inc., 1970.

Los Alamos Historical Society, *When Los Alamos Was a Ranch School*. Santa Fe, N.M.: Sleeping Fox Enterprises, 1974.

McPhee, John, *The Curve of Binding Energy*. New York: Farrar, Straus & Giroux, Inc., 1974.

Michelmore, Peter, *The Swift Years: The Robert Oppenheimer Story*. New York: Dodd, Mead & Co., 1969.

Moss, Norman, *Men Who Play God: The Story of the Hydrogen Bomb*. New York: Harper & Row, Publishers, 1969.

Pacific War Research Society, *The Day Man Lost: Hiroshima 6 August 1945*. Palo Alto, Ca: Kodansha International/USA, 1972.

Pettitt, Roland A., *Los Alamos Before the Dawn.* Los Alamos, N.M.: Pajarito Publications, 1972.

Rabi, Isidor, et al., *Oppenheimer.* New York: Charles Scribner's Sons, 1969.

Sherwin, Martin J., *A World Destroyed: The Atomic Bomb and the Grand Alliance.* New York: Alfred A. Knopf, Inc., 1975.

Smyth, Henry D., *Atomic Energy for Military Purposes.* Princeton, N.J.: Princeton University Press, 1947.

Stern, Phillip M., *The Oppenheimer Case: Security on Trial.* New York: Harper & Row, Publishers, 1969.

Stimson, Henry L., and McGeorge Bundy, *On Active Service in Peace and War.* New York: Harper & Row, Publishers, 1948.

Thomas, Gordon and Max Morgan Witts, *Enola Gay.* New York: Stein & Day Publishers, 1977.

Ulam, Stan M., *Adventures of a Mathematician.* New York: Charles Scribner's Sons, 1976.

U.S. Atomic Energy Commission, *In the Matter of J. Robert Oppenheimer.* Washington, D.C.: Government Printing Office, 1954.

Wohlberg, Margaret, *A Los Alamos Reader: 1200 A.D. to Today.* Los Alamos, N.M.: Los Alamos County Museum of History, 1976.

Works Progress Administration, *New Mexico: A Guide to the Colorful State.* Albuquerque, N.M.: University of New Mexico Press, 1945.

York, Herbert, *The Advisors: Oppenheimer, Teller and the Superbomb.* San Francisco: W. H. Freeman and Co., Publishers, 1976.

GLOSSARY

Absorption: The process by which the number of particles entering a body of matter is reduced by interaction of the particles with the matter; not to be confused with the concept of "capture."

Accelerator: A device or machine for increasing the velocity and energy of charged elementary particles. Types of accelerators include Cockcroft-Walton accelerators and Van de Graaff generators.

Atom: A particle of matter indivisible by chemical means and the fundamental building block of the chemical elements.

Atomic Bomb: A bomb whose energy comes from the fission of heavy elements, such as plutonium or uranium.

Atomic Number: The number of protons in the nucleus of the atom and its positive charge.

---

*Glossary adapted from Nuclear Terms (Atomic Energy Commission, n.d.)

Atomic Weight: The mass of an atom relative to other atoms; it is approximately equal to the total number of protons and neutrons in the nucleus.

Background Radiation: The radiation in man's natural environment, including cosmic rays and radiation from the naturally radioactive elements, both inside and outside the bodies of men and animals.

Blast Wave: A pulse of air, propagated from an explosion, in which the pressure increases sharply at the front of a moving air mass.

Capture: A process in which an atomic or nuclear system acquires an additional particle.

Chain Reaction: A reaction that stimulates its own repetition. In a fission chain reaction a fissionable nucleus absorbs a neutron and fissions, releasing additional neutrons. These in turn can be absorbed by other fissionable nuclei, releasing still more neutrons. A fission chain reaction is self-sustaining when the number of neutrons released in a given time equals or exceeds the number of neutrons lost by absorption in nonfissioning material or by escape from the system.

Cosmic Rays: Radiation from many sources, but primarily from atomic nuclei of very high energies originating outside of the earth's atmosphere.

Critical: Capable of sustaining a chain reaction.

Critical Assembly: An assembly of sufficient fissionable material and moderator to sustain a fission chain reaction at a very low power level.

Critical Experiment: An experiment to verify or supplement calculations of the critical size and other physical data affecting a reactor design or, in some cases, a bomb design.

Critical Mass: The smallest mass of fissionable material that will support a chain reaction.

Criticality: The state of a nuclear bomb core (or reactor core) when it is sustaining a chain reaction.

Cross Section: A measure of the probability that a nuclear reaction will occur. It is the apparent (or effective) area presented by a target nucleus (or particle) to an oncoming particle.

Curie: The basic unit to describe the intensity of radioactivity. Named for Marie and Pierre Curie, who discovered radium in 1898.

Delayed Neutrons: Neutrons emitted by radioactive fission products in a bomb or reactor over a period of seconds after a fission takes place. Fewer than one percent of the neutrons are delayed, the majority being "prompt" neutrons.

Electron: An elementary particle with a unit negative electrical charge and a mass 1/1837 that of the proton. Electrons surround the positively charged nucleus and determine the chemical properties of the atom.

Fast Neutron: A neutron with energy greater than approximately 100,000 electron volts.

Fireball: The luminous ball of hot gases that forms a few millionths of a second after a nuclear explosion.

Fission: The splitting of a heavy nucleus into two approximately equal parts—which are nuclei of lighter elements—accompanied by the release of a relatively large amount of energy and generally one or more neutrons.

Fissionable Material: Commonly used as a synonym for fissile material and includes material that can be fissioned by fast neutrons such as Uranium 238 and Plutonium 239.

Gamma Rays: High-energy, shortwave-length electromagnetic radiation that is very penetrating and fatal to the human body in sufficient doses. Gamma rays always accompany fission.

Heavy Water: Water containing significantly more than the natural proportion of heavy hydrogen (deuterium) atoms (more than one in 6500) to ordinary hydrogen atoms. Heavy water is used as a moderator in some reactors.

Hydrogen Bomb: A nuclear weapon that derives its energy largely from fusion.

Implosion Weapon: A weapon in which a quantity of fissionable material, less than a critical mass at ordinary pressure, has its volume suddenly reduced by compression (through use of explosives) so that it becomes supercritical and produces an explosion.

Isotope: One of two or more atoms with the same atomic number (the same chemical element) but with different atomic weights.

Kiloton Energy: The energy of a nuclear explosion that is equivalent to that of an explosion of 1,000 tons of TNT.

Mass: The quantity of matter in a body, and often used as a synonym for weight.

Moderator: A material, such as ordinary water, heavy water, or graphite, used in a reactor to slow down high-velocity neutrons, thus increasing the likelihood of further fission.

Natural Uranium: Uranium as found in nature, containing 99.3 percent of Uranium 238, .7 percent of Uranium 235, and a trace of Uranium 234.

Neutron: An uncharged elementary particle with a mass slightly greater than that of the proton, and found in the nucleus of every atom heavier than hydrogen.

Neutron Capture: The process in which an atomic nucleus absorbs or captures a neutron. The probability that a given material will capture a neutron is measured in terms of its cross section.

Nuclear Reaction: A reaction involving a change in an atomic nucleus such as in fission, fusion, neutron capture, or radioactive decay. A chemical reaction involves only the electron structure surrounding the nucleus.

Nucleus: The small, positively charged core of an atom. It is only 1/10,000 the diameter of the atom, but contains nearly all the atom's mass.

Particle: A minute constituent of matter, generally one with a measurable mass. The primary particles involved in radioactivity are alpha particles, beta particles, neutrons, and protons.

Pile: An old term for nuclear reactor. The name came from Fermi's experiments in building a reactor from piling up graphite blocks and natural uranium.

Plutonium: A heavy, radioactive, and manmade metallic element with an atomic number of 94. It is used primarily as reactor fuel and in weapons.

Prompt Neutrons: Neutrons that are emitted immediately following nuclear fission.

Proton: An elementary particle with a single positive charge and a mass approximately 1,837 times that of an electron.

Radioactivity: The spontaneous decay or disintegration of an unstable atomic nucleus usually accompanied by the emission of ionizing radiation.

Reflector: A layer of material immediately surrounding a bomb's nuclear core (or around a reactor's core) which scatters back or reflects into the core many neutrons that would otherwise escape. Also called a tamper.

Scattering: A process that changes a particle's trajectory.

Shock Wave: A pressure pulse in air, water, or earth, propagated from an explosion.

Slow Neutron: A thermal neutron.

Subcritical Assembly: A bomb core, or reactor core, that contains a mass of fissionable material and moderator whose effective multiplication factor is less than one and that hence cannot sustain a chain reaction.

Subcritical Mass: An amount of fissionable material that is insufficient in quantity or of improper shape to sustain a fission chain reaction.

Supercritical Mass: A mass of nuclear material in which each generation of neutrons is greater than the previous generation.

Thermal Neutron: A neutron in thermal equilibrium with its surrounding medium; these are neutrons which have been slowed down by a moderator.

Thermonuclear Reaction: A reaction in which very high temperatures bring about the fusion of two light nuclei to form the nucleus of a heavier atom, releasing a large amount of energy. In a hydrogen bomb, the high temperature needed to initiate the thermonuclear reaction is produced by using a preliminary fission bomb.

Uranium: A radioactive element with the atomic number 92 and an average atomic weight of 238. One of two isotopes is Uranium 235, which is fissionable.

Yield: The total energy released in a nuclear explosion. It is usually expressed in terms of TNT equivalents (the quantity of TNT needed to produce a similar amount of nuclear energy). Low-yield weapons are those less than 20 kilotons of TNT. Intermediate-yield weapons produce from 200 kilotons to 1 megaton.

INDEX

Abelson, Phillip, 21
Ackerman, Jim, 126
Administrative Board, Los Alamos Laboratory, 111, 114, 115, 189
  See also Governing Board
Advisory Committee on Uranium, 24, 25, 26, 28
Agnew, Harold, 59, 63, 176, 177, 184
Alamogordo Bombing Range, 7, 147, 161, 171
Alberta Project, 113, 117, 122, 133, 134, 174, 175, 176, 198
Albuquerque, New Mexico, 7, 17, 43
Albuquerque Engineering District, 43
Allison, Sam, 28, 30, 112, 133, 135, 168, 188
Alpha Site, 124
ALSOS mission, 142
Alvarez, Luis, 62, 87, 111, 126, 133, 135, 162, 176, 177, 208, 211
Amos, 123
Anchor Ranch, 82
Anchor Site, 132
Anderson, Carl, 40
Anderson, Herbert, 145, 161, 162
Anderson, John, 148
April Reviewing Committee, 79
Archie device, 83, 123
Army, U.S., 8, 16, 29, 32, 33, 34, 42, 65, 71, 80, 98, 136, 177
Arnold, Henry, 183
Ashbridge, Whitney, 111, 114
Ashley Pond, 44
Ashworth, Frederick, 175, 185–186
Associated Press (AP), 170–171
Atom smashers, *see* Particle accelerators
Atomic bomb, 3, 4, 16, 49
  assembly methods, 51–53, 121
  chain reaction, 5, 21, 22, 25, 38, 49, 50–51, 74
  critical mass, 50, 51–52, 77, 118, 124, 132
  cross sections, 49–50
  detonators, 126, 133, 135
  explosive lenses, 126, 132, 133, 135, 138, 160, 163, 177
  first explosion of, 168–170
  fission research, 47–53
  future discussed, 191–192
  initiator, 118–119, 135
  neutrons, 20, 50–51, 61
  as political weapon, 143, 172
  tamper, 50, 78, 163

Atomic bomb (*cont.*)
  See also Fat Man, Gun method, Implosion method, Little Boy
Atomic Bomb Group, Wendover Field, 175
Atomic Energy Commission (AEC), 209, 210, 211
Atomic pile, 25, 32, 37–38
Autocatalytic process, 52, 121

Bacher, Jean, 101
Bacher, Robert, 60, 67, 68, 69, 77, 78, 79, 81, 110, 111, 112, 113, 119, 124, 126, 132, 134, 162, 188, 208
Bainbridge, Kenneth, 8, 28, 62, 63, 68, 81, 87, 111, 112, 126, 144–145, 146–151, 152, 154, 159, 160, 164, 166, 167, 168, 169–170, 171, 183
Bainbridge, Peg, 101
Bandelier, Adolph, 14
Bandelier National Monument, 19
Bard, Ralph, 181
Barnett, Henry, 171
Base Camp, *see* Trinity Test
Bayo Canyon, 127
Beahan, Kermit, 185–186
Belcher, Phil, 171
Berkeley conferences, 36, 39, 42
Beta Site, 124
Betatron, 132–133
Bethe, Hans, 6, 36, 59, 60, 61, 63, 68, 69, 76, 77, 85, 88, 109, 111, 118, 119, 120, 131, 135, 150, 196, 208, 211
Bhagavad-Gita, 134, 170
Birch, A. Francis, 113, 133
Bock's Car, 185–186
Bohr, Niels, 22, 59, 60, 61, 99, 112, 143, 182, 192, 208
Born, Max, 35
Bradbury, Norris, 60, 62–63, 112, 113, 126, 133, 149, 160, 163, 164, 175, 178, 198–202, 203, 208, 211
Breit, Gregory, 34, 36, 42
Briggs, Lyman J., 24, 25, 26, 27, 28, 30, 31, 33
Brixner, Berlyn, 63
Brode, Bernice, 93, 101
Brode, Robert B., 81, 82–83, 93, 113, 123, 133
BRONX project, 177
B-29 (airplane), 83–84, 136, 167, 175
Bulletin (newspaper), 101

229

230/ Index

Bundy, Harvey H., 181
Burke, W. E., 156
Bush, Harold, 148, 166, 168
Bush, Vannevar, 8, 25, 27, 28, 29, 30, 31, 32–33, 34, 37, 39, 42, 48, 69–70, 81, 114, 129, 157, 165, 168, 171, 181, 191–192, 200
Byrnes, James F., 129, 182

California Institute of Technology, 35, 114
CAMEL project, 114, 133
Carlson, Robert, 127, 149
Carrizozo, New Mexico, 168
Chadwick, James, 20, 37, 59, 61, 102, 111, 165, 168
Chadwick, Lady, 102
Chain reaction, 5, 21, 22, 25, 38, 49, 50–51, 74
Chapin, Roy, Jr., 15
Chemistry and Metallurgy Division, Los Alamos Laboratory, 68, 79, 118, 127, 134, 159
Christy, Robert, 78, 79, 150
Churchill, Winston, 9, 25, 60, 61, 143, 173, 192
Clayton, William, 181
Clinton Reactor, Washington, 80, 86, 88, 127, 132
Cockcroft-Walton accelerator, 44, 78, 161
Colloquium, 69
Combined Policy Committee, 61
Compañia Hill, 165, 166, 167, 168, 169, 170, 172
Compton, Arthur, 181, 194
Compton, Karl T., 24, 25, 28, 29, 30, 33, 34, 36, 37, 38, 40, 45, 85, 157, 159, 181, 191–192
Conant, James B., 8, 25, 27, 28, 29, 31, 33, 34, 37, 38, 42, 48, 60, 63, 64–65, 67, 81, 86, 88, 109, 114, 133, 157, 165, 167, 171, 181, 200
Condon, Edward, 28, 66–67, 68, 71
Connell, Arthur J., 15, 18
Coronado, Francisco Vasquez de, 13
Cowpuncher Committee, Los Alamos Laboratory, 112–113, 114, 117, 135, 136, 151
Critchfield, Charles, 81
Critical Assemblies team, 124, 135
Critical mass, 50, 51–52, 77, 118, 124, 132
Crosby, John, 15
Cross sections, 49–50
Curie, Marie, 79
Curie, Pierre, 79
Cyclotron, 7, 21, 26, 35, 78
Cyclotron group, 77

Daghlian, Harry, 125
Delight Makers, The (Bandelier), 14
Delivery group, 83–84, 175
Democritus, 20
de Silva, Peer, 98, 99, 147, 152, 189
Destination, *see* Tinian Island
Detonators, 126, 133, 135
Deuterium, 48, 53, 77
Diffusion experiments, 76
Diffusion group, 118
Dill, John, 142
Doan, Richard, 30, 63

Dow, David, 111–112
DP Site, 129, 134
Dragon experiment, 124–125, 135
Dudley, William, 10
Dunning, John, 23, 24
DuPont, 37

Eareckson, William, 171
Einstein, Albert, 6, 23–24, 192
Eisenhower, Dwight D., 112, 142, 209, 211
Electromagnetic separation process, 26, 27, 32, 87
Electronics group, 77
Electrostatic generator, 21
Electrostatic generator group, 77
Element 93 (neptunium), 27
Element 94 (plutonium), 27–28
  See also Plutonium
"Encyclopedia," 196
Engel, Albert, 129
Enola Gay, 183, 184
Experimental Physics Division, Los Alamos Laboratory, 68, 77, 110
Explosives Division, Los Alamos Laboratory, 110, 122, 163

Farrell, Thomas, 8, 160, 164, 167, 168, 169, 177, 178, 181, 185
Fat Man, 3–4, 7, 8, 9, 63, 102, 113, 114, 117, 118, 119, 122, 123, 124, 126, 127, 128, 133, 134, 135, 136, 137, 143, 144, 149, 151, 160, 164, 168, 172, 173, 179, 182, 183, 185, 186, 187, 195, 201, 207
  See also Atomic bomb, Implosion method
Ferebee, Thomas, 184
Fermi, Enrico, 6, 22, 23, 24, 25, 28, 30, 37–38, 39, 59, 60, 61, 63, 74, 77, 79, 85, 88, 99, 110, 111, 119–120, 122, 124, 131, 149, 168, 169, 172, 173, 181, 188, 192, 194, 195, 207, 208, 210–211
Fermi Award, 210
Feynman, Richard, 63, 76, 99, 168
Fifth Washington Conference on Theoretical Physics, 22
Fission, 21, 22, 23, 42, 48, 49–50, 60, 75, 122, 138
509th Composite Group, 175, 176, 177
Flanders, David, 76
Fowler, Gene, 113
Franck, James, 59
Frankel, Stanley, 59
Frey, Evelyn, 14
Frey, George, 14
Frijoles Canyon, 14
Frisch, Otto, 22, 60, 61, 124, 125, 134–135
Froman, Darol K., 62, 77, 132, 161, 196
Fuchs, Klaus, 61–62, 168, 173, 211–212
Fuller Lodge, 44
Fussell, Lewis, 113, 133, 145, 148
Fuze Development group, 82–83

Gadget Division, Los Alamos Laboratory, 119, 124, 132, 163

Index /231

Gamma rays, 161, 172–173
Garner, Charles, 128
Gaseous diffusion process, 26, 34
Germany, 141, 142
Gold, Harry, 62
Goudsmit, Samuel, 142
Governing Board, Los Alamos Laboratory, 68, 69, 71, 76, 77, 81, 85, 86, 87, 98, 109, 111, 120, 144–145, 147, 149
Grand Artiste, 185
Graves, Alvin, 62, 168
Graves, Elizabeth, 168
Great Britain, involvement in bomb development, 60–61
Green, Priscilla, 45
Green Hornet, 184
Greenglass, David, 62, 173, 211, 212
Greenglass, Ruth, 62
Greisen, Kenneth, 161, 164
Ground Zero, see Trinity Test
Groves, Leslie R., 6, 7, 8, 11, 15, 16, 37, 38–39, 40, 41, 43, 45, 47, 48, 55, 59, 60, 64–65, 66, 67, 69, 70, 71–73, 74, 75, 76, 77, 78, 80, 81, 86, 87, 88, 89, 92, 93, 94, 95, 96, 97, 98, 100, 111, 114, 115, 116–117, 128, 129, 133, 138, 145, 146, 147, 148, 149, 151, 152, 153, 155, 157, 161, 162, 164, 165, 166, 167, 168, 170–171, 172–173, 175, 176, 177, 178, 181, 182, 183, 187, 189, 190, 193, 194, 195, 196, 197, 198, 199, 201, 202, 207, 211
  appointed director of Manhattan Project, 5–6, 10, 34
  character and description, 5, 34
  decision-making ability, 41–42
  determination to use atomic bomb, 142–143
  historical legacy, concern about, 143
  opposition to Fermi, 120
  relationship with Oppenheimer, 58
  selects Los Alamos site, 16–19
  selects Oppenheimer as Director of Los Alamos Laboratory, 40–42
  views on control of atomic weapons, 191
Guadalupe-Hidalgo, Treaty of, 14
Gun Group, Los Alamos Laboratory, 117
Gun method, 51–52, 53, 75, 76, 78, 81–82, 85, 86, 87, 112, 133

Hahn, Otto, 21–22
Hanford plant, Washington, 49, 75, 78, 80, 129, 134, 151, 197
Harrison, Katherine, see Oppenheimer, Katherine
Harrison, Richard, 35
Hawkins, David, 68
Heisenberg, Werner, 27
Hempelmann, Louis, 80–81, 153, 176
Henderson, Robert, 147
Heydenburg, Norman, 36
High Explosives Group, Los Alamos Laboratory, 144
Higinbotham, Willie, 132
Hillhouse, Dorothy, 101

Hiroshima, 21, 186, 187, 191, 194, 207
  bombing of, 184–185
  selected as target, 179, 180, 181, 183, 184
Hirschfelder, Joe, 149
Hitler, Adolph, 142
Holloway, Marshall, 8, 59, 133, 162, 163, 168
Hooper, Stanford, 23
Hornig, Donald, 8, 63, 145, 166, 168, 176, 208
Hubbard, John, 151, 154, 166, 167
Hughes, A. L., 68, 72
Hydrogen bomb, 39, 48, 53–54, 55, 76, 77, 120–121, 137–138, 201–202, 210–211
  early discussion of, 37
  Oppenheimer's views on, 137–138
  use of deuterium, 121

"Immediate Aftereffects of the Gadget" (Bethe and Christy), 150
Implosion method, 52–53, 75, 76, 81, 84–85, 86, 87–89, 109, 112, 113, 114, 117, 118, 119, 122, 126, 131–132, 133, 136, 138, 141, 144, 161
Indianapolis, U.S.S., 177
Inglis, Betty, 101
Initiator, 118–119, 135
Institute for Advanced Study, Princeton, 210
Instrumentation ("Detector") group, 77–78
International Scheduling Conference, 112
Isotron Project, 60, 68

Japan, war against, 141, 142, 144, 172, 173, 174–187 *passim*
Jemez Mountains, 4, 10, 11, 12, 46, 91
Jemez Springs, 10, 17
Jette, Eleanor, 101
Jette, Eric, 62, 136
Jewett, Frank B., 25
Johnson, Lyndon B., 210
Joliot, Frédéric, 23, 85
Jornado del Muerte (Dead Man's Route), 7, 147, 158
Jumbo, 127, 145, 150, 172

Kaiser Wilhelm Institute for Chemistry, 21
Kearny, Stephen W., 14
Kennedy, John F., 210, 211, 212
Kennedy, Joseph, 27, 59, 60, 62, 68, 79, 80–81, 111, 127, 128, 196
Kingman, see Wendover Field
Kirtland Air Base, Albuquerque, 167
Kistiakowsky, George, 26, 60, 61, 63, 68, 81, 87, 88, 110, 111, 112, 114, 117, 122, 126, 131, 144, 145, 148, 150, 160, 163, 164, 166, 167, 168, 170, 176, 178, 187, 195, 196, 208
Kokura, Japan, 180, 181, 183, 184, 186
Konopinski, Emil J., 77
Krohn, Bob, 63, 168
K Site, 132
Kyoto, Japan, 180, 181

Laboratory Coordinating Council, 69
Lamy, New Mexico, 90
Landsdale, John, 41

232/ Index

Landshoff, Rolf, 61
Lapaca, Mary, 170
Lauritsen, Charles C., 112, 114, 133, 165
Lawrence, Ernest O., 7–8, 21, 26, 27, 30, 31, 33, 35, 40, 45, 55, 59, 66, 87, 100, 157, 159, 165, 181, 192, 194, 195
Lawrence, William, 165, 170
Le May, Curtis, 184
Lenses, explosive, 126, 132, 133, 135, 138, 160, 163, 177
Lewis, W. K., 48
Liquid thermal diffusion process, 26
Little Boy, 122, 123, 128, 133, 134, 136, 143, 144, 172, 179, 182, 183, 184, 185, 207
  See also Atomic bomb, uranium gun
Livermore Laboratory, 210
Lofgren, Edward, 135
Los Alamos, New Mexico, 4, 11, 17, 90, 212
Los Alamos Laboratory, 4, 6, 9, 43–56 passim, 64, 74, 116, 130, 134, 144, 175, 187, 188
  administrative organization, 66–73
  housing, 45, 72, 93–95, 115
  living conditions, 90–106, 115
  physical setting, 44–46, 97, 103–104
  postwar confusion and discontent, 194–197
  postwar conversion, 188–203 passim
  recruiting for, 97, 113–115
  recruits personnel for bombing of Japan, 176–177
  retrospective view of, 207–209
  Review Committee, 48, 54, 55, 66
  salaries, 72, 113
  security, 69–70, 93, 97–100, 165–166, 194
  selected as site for weapons research, 16–19
  social life at, 101–106
  work day, 100
Los Alamos Ranch School for Boys, 14–15, 43, 46, 91, 94

McDaniel, Boyce, 163–164
McDonald ranch house, 160, 162, 213
Mack, Julian, 145, 149, 161–162, 168
McKee Company, 94
McKibben, Dorothy, 46, 91, 93
McKibben, Joseph, 36, 164, 166, 167, 169
McMillan, Edwin, 27, 44, 59, 68, 81, 111, 117, 118, 132
Manhattan Engineering District (Manhattan Project), 4, 5, 8, 9, 10, 16, 19, 33, 34, 37, 39, 41, 43, 55, 61, 65, 70, 74, 116, 128, 129, 134, 143, 144, 151, 178–179, 182, 191, 192, 193, 195, 197, 211
Manley, John, 36, 44, 45, 47, 60, 77, 78–79, 94, 97, 119, 149, 161, 168, 194 211
Manley, Kay, 101
Mark, Carson, 61
MARK I, 116–117
MARK II, 116–117
MARK III, 116–117
Marshall, George, 5, 29, 32, 33, 34, 142, 143, 181, 187
Marshall, James C., 33, 34
Marshall, Sam, 128

Martin Aircraft Plant, 175
Matthias, Franklin, 129
May-Johnson Bill, 195, 199
Meitner, Lise, 22, 60
Memorandum on the Los Alamos Project, 92–93
Metal purification, 80
Metallurgical Laboratory, University of Chicago, 28, 30, 36, 39, 40, 68, 70, 80, 112, 114, 191, 192, 193
  creates first controlled chain reaction, 37–38
Miera, Jose, 152, 168
Military Policy Committee, 34, 39, 42, 55, 69
Mitchell, Dana, 68, 111
Moon, Phillip, 61, 145
Morgan Company, 94
Morrison, Philip, 133–134
Morrison, Sam, 113
Muroc Army Air Force Base, Utah, 83
Murphee, Edgar, 30, 31, 33

Nagasaki, Japan, 21, 179, 183, 186, 187, 191, 207
National Defense Research Committee (NDRC), 24–25, 26, 47–48, 110
Neddermeyer, Seth, 52–53, 63, 76, 81, 85, 86, 87, 88, 111, 132, 150
Nelson, Eldred, 59
Neptunium (Element 93), 27
Neutron, 20, 50–51, 61
Neutron Source Research group, 77
Nichols, Kenneth, 33, 88, 129
Niigata, Japan, 180, 181, 183
Nolan, James, 176
Norstad, Lauris, 178
North 10,000 Shelter, 159, 168, 171
Nuclear efficiency, 76
Nuclear Test Ban Treaty (1963), 210

Oak Ridge Laboratory, Tennessee, 17, 34, 49, 63, 70, 114, 129, 134, 197
Office for Emergency Management, 28
Office of Scientific Research and Development (OSRD), 8, 28, 33, 39, 72
  See also S-1 Committee
O'Leary, Jean, 172
Omega Site, 79, 124, 135
Oñate, Don Juan, 13
109 Palace Street, Santa Fe, 91, 93
Oppenheimer, Frank, 35, 41, 63, 149
Oppenheimer, J. Robert, 6, 7, 8, 10, 11, 15, 16, 17, 19, 34, 35–36, 37, 39, 42, 43, 44, 45, 47, 48, 49, 50, 52, 54, 56, 60, 61, 62, 63, 64, 67, 69, 70, 71–73, 74, 75, 77, 80, 81, 85, 86, 87–89, 91, 93, 94, 96, 97, 99, 100, 101–102, 105, 114, 116, 117, 118, 119–120, 124, 126, 127, 128, 130, 133, 134, 141, 142, 144–145, 146, 147, 148, 150, 151, 154, 155, 157, 159, 161, 164–165, 167, 168, 169, 170, 171, 172, 176, 177, 178, 179, 181, 182–183, 186–187, 188–189, 192, 193, 194, 200, 201, 202, 207, 208, 213
  appointed Director of Los Alamos Laboratory, 40–42

Oppenheimer, J. Robert (cont.)
    assesses Los Alamos Laboratory, 136–138
    assessment of, 209–210
    character and description, 5
    consolidation of uranium projects urged by, 16, 39–40
    early involvement in fission research, 36
    early life, 4, 35, 40–41
    enters bomb movement, 29
    hydrogen bomb, view of, 137–138, 209
    leadership of Los Alamos Laboratory, 57–59
    loyalty questioned, 41
    organizes administration of Los Alamos Laboratory, 66–73
    relationship with Groves, 58
    reorganizes Los Alamos Laboratory, 109–115, 116–122
    resigns from Los Alamos Laboratory, 198–199, 202–203
    responsibilities at Los Alamos outlined, 64–66
    role in postwar activities of Los Alamos Laboratory, 190–191, 195, 197
    declared security risk, 209–210
    teaching years, 36
Oppenhiemer (Harrison), Katherine, 35–36, 41, 101, 105
Oppenheimer, Peter, 94
Oppenheimer, Toni, 94
Ordnance Division, Los Alamos Laboratory, 68, 81, 82, 85, 86, 87, 110, 122, 123, 126, 174–175
Osborn, David, 15
Oscurro Mountains, 7

Page, Arthur, 181
Pajarito Plateau, 4, 11, 12, 20, 43, 45, 89, 212
Palmer, T. O., 153
Parsons, William, 60, 63, 68, 71, 76, 81, 82, 85, 86, 87, 94, 100, 111, 112, 113, 122, 126, 131, 133, 136, 147, 162, 167, 174–175, 176, 177, 178, 181, 183, 184, 185, 186, 188, 195, 207
Particle accelerators, 26, 50
Pash, Boris T., 41, 142
Pegram, George, 23, 24, 28
Peierls, Genia, 102
Peierls, Rudolph E., 61, 84, 102, 118, 132
Pelly, Lancelot Inglesby, 15
Penney, William, 61, 161, 176, 179
Peralta, Don Pedro, 91
Plutonium 239 (Pu 239), 4, 5–6, 9, 21, 28–29, 30, 32, 37, 39, 48, 49, 50, 51, 52, 53, 54, 55, 74, 75, 77, 78, 79, 80–81, 82, 85–86, 88, 109, 116, 117, 124, 125, 127, 128, 131, 134, 136, 138, 141, 143, 146, 159–160, 161, 162–163, 177, 197
Plutonium Group, 128
Plutonium purification process, 80
Polonium, 85, 124, 128
Polonium Group, 128
Pond, Ashley, 14–15
Potsdam Conference, 151, 157, 168, 172
Project Y, see Los Alamos Laboratory

Projectile and Target Group, 81
P Site, 124
Purnell, William, 34

Quebec Agreement, 143

Rabi, Isidor I., 22, 59, 61, 67, 111, 112, 149, 165, 168
Radio Corporation of America (RCA), 83
Radio Saipan, 185
Radioactive contamination, 178
Radioactivity group, 77
Radiolanthum, 132
Ramsey, Norman, 62, 83–84, 111, 113, 147, 175, 176, 177, 181, 183
Reed, John J., 15
Research Division, Los Alamos Laboratory, 110, 119
    See also Experimental Physics Division
Review Committee, 134
Richards, Hugh, 161
Richardson, Vaughn, 159, 160
Rodriguez, Augustin, 13
Roosevelt, Franklin D., 6, 9, 23–24, 25, 28, 29, 30, 32, 33, 39, 60, 61, 69, 75, 76, 89, 117, 129, 134, 143, 191, 192
Rose, E. L., 48
Rosenberg, Ethel, 62, 212
Rosenberg, Julius, 62, 212
Rossi, Bruno, 61, 78, 132, 133, 161
Rowe, Hartley, 112, 113
Roy's Café, 152
Rufus, 125

Sachs, Alexander, 24
San Idelfonso, New Mexico, 13, 104
Santa Fe, New Mexico, 4, 7, 45–46, 47, 64, 90–91, 105–106
Schreiber, Raemer, 63, 176–177
Science Advisory Council, 24
Scientific Panel, 193
Seabees, U.S. Navy, 176
Seaborg, Glenn, 27–28, 68, 78
Segré, Emilio, 27, 28, 59, 61, 77, 85, 119, 161
Serber, Robert, 36, 47, 48, 49–51, 52, 54, 60, 76, 118, 176
Shane, C. D., 111, 113
Shapiro, M. M., 123
Slotin, Louis, 125, 134–135, 160
Smith, Alice Kimball, 101, 208
Smith, Cyril, 111, 196
Smith, Henry D., 28
Smith, Lloyd, 28
Socorro, New Mexico, 158, 165, 170
Somervell, Brehon B., 34
S-1 Committee, 28, 30, 32, 33, 34, 181–182
South 10,000 Shelter, 7, 8, 159, 166, 167, 168, 169, 171
S Site, 62, 124, 126, 133, 135, 160
Stalin, Josef, 9, 138, 143, 157, 172, 173
Stearns, Joyce, 179, 180
Stevens, W. A., 126

Stimson, Henry L., 18, 29, 116, 117, 129, 138, 142, 143, 144, 157, 172, 181, 182, 183, 187, 191, 192, 193, 194
Stone and Webster Company, 37
Strassmann, Fritz, 21
Styer, Wilhelm D., 32, 33, 34
Sundt Company, 94
Super and General Theory Group, Los Alamos Laboratory, 120–121
Superbomb, see Hydrogen bomb
Sweeney, Charles, 185
Symmetrical explosives, 135
Szilard, Leo, 6, 23, 24, 143, 192–193

Tamper, see Atomic bomb
Target Committee, 178–181
Tatlock, Jean, 41
Taylor, Geoffrey, 59, 61, 165, 168
Technical Area, Los Alamos Laboratory, 44, 45, 46, 130, 134
Technical Board, Los Alamos Laboratory, 111, 112
See also Governing Board
Technical and Scheduling Conference, 112, 133, 135
Teller, Edward, 6, 24, 36, 53, 54, 59, 60, 61, 76, 77, 82, 84, 85, 99, 111, 120–121, 137–138, 168, 193, 202, 210
Theoretical Division, Los Alamos Laboratory, 44, 54, 68, 76, 84, 86, 118, 124
Thermonuclear bomb, see Hydrogen bomb
Thomas, Charles, 80, 88, 165
Thompson, Roy, 163
Tibbets, Paul, 183, 184
Tibbits, William, 175, 177
Tilano, 91, 104–105
Tinian Island, 136, 175–176, 177, 179, 183, 184, 185, 187
Titterton, Ernie, 61, 161
Tolman, Richard, 47, 48, 51, 52, 75, 77, 85, 120, 157, 164, 165, 172–173, 189
Trinity Test (TR), 7, 8, 9, 102, 113, 146–173 passim, 177, 178, 182, 183, 184, 191, 194, 198, 201, 207, 212
  accidental bombing of, 161
  Base Camp, 7, 8, 148, 151, 159, 165, 168, 169, 212
  Ground Zero, 7, 8, 148, 150, 159, 161, 162, 163, 164, 172, 212
  preliminary test held, 154
  security precautions, 152–153
  selection of name, 7, 46
  selection of site, 146–147
  weather, 166, 167–168
  Zero Hour, 7, 166, 167, 168
Tritium, 77
Truman, Harry S, 8, 9, 138, 144, 151, 157, 165, 168, 172, 173, 182, 183, 187, 191, 195, 200
"Tube Alloys" project, 60, 61
Tuck, James, 61, 84–85, 149, 199
240 isotope, 119

Ulam, Stanislav, 61

Underhill, Robert, 45
University of California, 43–44, 72, 176
University of California Radiation Laboratory, 27
Uranium 235, 5–6, 9, 21, 23, 24, 25–26, 27, 28, 29, 30, 32, 37, 39, 48–49, 50, 51, 52, 53, 54, 55, 70, 75, 76, 77, 78, 79, 80, 82, 86, 109, 112, 116, 117, 122, 124, 127, 128, 131, 134, 136, 138, 141, 143, 177, 197
Uranium 238, 4, 21, 23, 25–26, 27, 50, 87
Uranium Committee, see Advisory Committee on Uranium
Uranium gun, 109, 117, 118, 122, 131, 134, 138
  See also Little Boy
Uranium hexafluoride, 26–27, 128
Uranium hydride, 55, 75
Uranium purification process, 80
Urey, Stan, 30, 33

Van Gemert, Robert, 155
Van de Graaff, Robert J., 21
Van de Graaff accelerator, 21, 44, 63, 78
Van Kirk, Ted, 184
Van Vleck, J. H., 48
Veeck, William, 15
Vidal, Gore, 15
Von Neumann, John, 59, 61, 86, 124, 132, 165, 179, 208, 211

Wahl, Arthur, 27
Wallace, Henry A., 29
War Production Board, 33, 34
Warner, Edith, 91, 104–105
Warner, Roger, 133
Water Boiler (Reactor) group, 78, 79, 119, 122, 124
Watson, Edwin, 24
Weapons Committee, Los Alamos Laboratory, 113, 175
Weapons Physics Division, Los Alamos Laboratory, 110, 122
Weisskopf, Victor, 61, 76, 149, 169
Wendover Field, Utah, 71, 134, 198
West 10,000 Shelter, 159, 166, 168
"What to Do Now" (Kistiakowsky), 196
Wigner, Eugene, 23, 24, 30
Williams, John, 36, 77, 78, 119, 149, 152, 157, 159, 167, 208
Wilson, E. B., 48
Wilson, Jane, 95
Wilson, Robert, 60, 68, 77, 78, 95, 110, 111, 119, 135, 161, 179, 196
Wood, Arthur, 15
World Destroyed, A (Sherwin), 182

X Site, 124
X-2 group, 144

Yokohama, Japan, 180
Y Site, 136

Z Division, Los Alamos Laboratory, 113
Zero Hour, see Trinity Test
Zinn, Walter, 38